Frames and Connections in the Governance of Global Communications

Frames and Connections in the Governance of Global Communications

A Network Study of the Internet Governance Forum

Elena Pavan

LEXINGTON BOOKS
Lanham • Boulder • New York • Toronto • Plymouth, UK

British Library Cataloguing in Publication Information Available

Library of Congress Cataloging-in-Publication Data

The hardback edition of this book was previously cataloged by the Library of Congress as follows:

Pavan, Elena,1980-
  Frames and connections in the governance of global communications : a network study of the Internet Governance Forum/ Elena Pavan.
     p. cm.
  Includes bibliographic references.
  1. Internet governance. 2. Communication—International cooperation.
  3. Telecommunications—International cooperation. I. Internet Governance Forum.
  II. Title.
   TK5105.8854.P38 2013
   384.3' 3—dc23
                            2011039646

  ISBN 978-0-7391-4643-9 (cloth : alk. paper)
  ISBN 978-0-7391-9059-3 (paperback)
  ISBN 978-0-7391-4645-3 (electronic)

This book is published with the contribution of the Department of Sociology and Social Research of the University of Trento and of the MIUR, in the context of the project MIUR-PRIN 2006 "Diritti di comunicazione fra locale e globale: reti di Movimento e trasformazioni dei processi di governance" (Contract 2006148230, University of Tranto and University of Padova).

To Annita, Carlo, Anna, and Vittoria,

    the cardinal points in my compass.

And to Thomas,

    my Northern star.

Contents

Acknowledgments

I owe particular thanks to my *maestro*, Mario Diani. Without his advice, his expertise, and his unique insights on society, my learning path would have been dry and dull.

I am profoundly indebted to Claudia Padovani, who has been following me in this long journey since I was just a neophyte, to both academia and life. She taught me the importance of communication for politics, for democracy, and for daily life. She is the researcher I wish I could become one day.

I am very grateful to GigaNet, the Global Internet Governance Academic Network, which welcomed me warmly when I was at the beginning of my path and has provided me with a comfortable and stimulating intellectual environment in which to grow. I would also like to express my most sincere acknowledgments to all colleagues, young and senior researchers, and to the patient professors who have discussed with me the different aspects of this work since its beginning.

Finally, I would like to acknowledge the support of the Department of Sociology and Social Research of the Università degli Studi di Trento in the realization of this book.

Introduction

This book stems from research conducted between 2006 and 2009 in the context of a national project titled "Communication Rights between Local and Global: Movement Networks and Transformations of Governance." The project was conducted in Italy through the joint work of two research teams from the Universities of Trento and Padova and started from an explicit acknowledgment of the increased relevance of global communications in the political agenda, not only in terms of tools and infrastructures but, more properly, as a field where relevant dynamics are taking place. Research activities revolved around one main research question—*How is Global Communication Governance* (GCG) *being structured?*—where GCG was understood, quite generally, as the sum of all political processes oriented toward the management of information and communication issues beyond the national level. In order to provide an answer, the project investigated two domains belonging to the broad GCG landscape and within which several interesting official and institutional processes were developing: the global Internet governance (IG) discussion and the discussion on media pluralism in the European Union (EU). Indeed, in that period, IG had emerged as the most relevant theme in the GCG field, thus involving institutional and non-institutional actors in an open and participatory confrontation, whereas the media pluralism case appeared to be particularly interesting for the study of multilevel, multiactor regional processes in the EU context.

The present work illustrates the main results coming from the examination of the IG case. As suggested by the title of the book, this study adopts a network approach to look at the transformations occurring within this specific domain, but, more generally, it also aims at stimulating a critical reflection on how to investigate the changes of global politics, giving specific attention to the participation of nonstate actors in official processes and to the multiplication of issues that are discussed when the focus is set on information and communication matters. As mentioned, both IG and media pluralism were selected as case studies, starting from the interesting features of the political processes that were developing in these domains. Indeed, an increased institutional attention (coming mainly from the United Nations [UN] in the IG case and the EU in the media pluralism case) led in both areas to the development of peculiar multiactor mechanisms aimed at clarifying their thematic boundaries and at defining the general roles and responsibilities of governmental and non-governmental actors in the steering and implementation of the politics that regulate them. Furthermore, both domains characterize themselves through their inherent relevance. In particular, IG constitutes one of the most important domains in the broader GCG landscape, one that captured global attention thanks to the mobilization of institutional and non-institutional actors within ad hoc generated spaces of debate such as the Internet Governance Forum (IGF). In emerging so remarkably on the global stage, discussions on IG ended up intertwining not only with other information and communication issues that were not previously referred to it (such as freedom of expression) but also with even broader conversations on transformations in supranational politics and on the modes of participation of non-institutional actors in them. Overall, then, the study of current IG debates has a twofold added value: it has a lot to say both on contemporary political global trends and on the more specific dynamics pertaining to the evolvement and management beyond the national level of information and communication issues.

However, this book does not assume that the political relevance of GCG and IG is self-evident or that all readers are familiar with these matters. Indeed, although GCG and IG are new, complex, and constantly evolving areas that are stimulating experimentation with political and research practices, they remain, in general, quite unfamiliar. The governance of information and communication issues fully permeates our lives, but we are seldom aware of this fact. Nowadays, Information and Communication Technologies (ICTs) are fundamental for creating, dissolving, and maintaining social ties among people, industries, organizations, and institutions worldwide. Communication flows are the very backbone of contemporary societies, and ICTs and Internet connections have a capillary and pervasive infrastructure that ends up having an actual influence on the way in which we perform our activities and pursue our long-term goals. Such a paramount role of global communications should illustrate the prominent position of information and communication governance matters in the political agenda. Because of the unprecedented embeddedness of ICTs in our lives, how these technologies are managed and how plans for their future developments are drafted will inevitably affect the way we live.

However, the popularity of GCG as a policy field is not that high yet. In fact, as we will see in more detail below, information and communication issues are rarely examined from a political perspective. For sure, there are important discussions on the *political use* of information and communications, but a more specific focus *on the processes* that regulate their development and the extension of their operational space is less often adopted. This paradox is most clear in the Internet case: we cannot even imagine ourselves without services built on top of the Internet, and our lives have certainly known an unprecedented improvement thanks to the acceleration of communication flows online. However, only a few of us are aware that what we can do online today is also the result of the complex governance mechanisms that regulate and guarantee the functioning of the Internet infrastructure but that also impact our "networked life" through the definition of constraints and possibilities in our online activity. Even fewer of us are aware that all Internet users are free to contribute to these governance practices or know that their involvement dates back to the very origins of Internet management, which have always been bottom-up and stemming from Internet users and uses.

One of the reasons why GCG and IG are not well known is the inherent complexity and multifaceted nature of these areas, whose management and study require specific and multiple competencies. Not only does the governance of global communications (and of the Internet in particular) involve dealing with many other issues, ranging from the technical developments of different media to users' protection and rights, but all these topics also have to be considered in relation to each other in a complex and dynamically evolving thematic maze that challenges the effectiveness and efficacy of existing governance arrangements. Therefore, to avoid getting lost after a few steps, practitioners and researchers who engage in these areas should be equipped with a certain sensibility for technical aspects and with a marked multidisciplinary attitude. However, this is easier said than done. A further aggravating element is that the study of GCG and its domains is confronted with an overall lack of sound and systematic theoretical and empirical frameworks for interpreting their complexity and setting up concrete research programs.

In our research project, we faced the challenges deriving from such a lack, and we elaborated an ad hoc approach to the study of governance dynamics of information and communication issues, using IG as a test case. This book is meant to offer a systematic account of the theoretical and empirical journey we engaged in to uncover the dynamics though which IG is consolidating at the global level as a relevant policy domain within the broader GCG landscape. It aims to do more than merely illustrate the results we derived from studying the IG case; it also retraces our conceptual and practical efforts to depict and systematically inquire into information and communication issues as relevant political matters on the global scale. Indeed, even before approaching the IG domain, we moved from the assumption that, in the governance of global communications, information and communication are not only *the means* through which processes are carried on but also *the very object of political confrontation*. Yet such a peculiar intertwinement of contents and processes has seldom been investigated in a systematic way. To fill in this lack from a theoretical perspective and in preparation for actual empirical investigation, we leaned on two main analytical tools, *frames* and *connections*, and we elaborated an analytic framework based on networks and whose validity should be extended to the whole GCG field. We then applied our framework to our study of the IG case and, in this way, both deepened our knowledge of this domain and started to compose the picture of how GCG is being structured. In this research context, we use *frames* to uncover the very *political substance* of IG and to analyze the various perceptions among institutional and non-institutional actors so as to discover the cognitive bases of possible political conflicts. Indeed, we believe that transformations in the ways of framing global issues (in our case, IG) are full of implications for the functioning and future development of all global policy domains. However, we also believe that it is in conjunction with *connections* that frames can actually influence how political dynamics are played out.

Here, we understand *connections* in a threefold way. In the first place, connections refer to the *interpersonal and cross-sector ties between institutional and non-institutional actors* that are particularly stimulated by the realization of participative official processes such as the IGF. As we will see below, since the IGF is meant to provide a space for fostering dialogue between actors, the political connections stimulated by it aim mainly at creating an overall collaborative environment where different political entities work together. However, this tendency toward cooperation does not eliminate the conflictual feature that generally characterizes the conduct of political dynamics. In fact, conflict is an inherent ingredient of multiactor dynamics in the IG domain, but it does not take the shape of physical confrontation. Rather, it is played out at the symbolic level and grows out of the different perceptions of the same issue that are held by different stakeholders. It then becomes crucial to understand connections also as *symbolic ties* that hold together and connect different conceptualizations of IG. Indeed, within an interactional environment like the one set up by the IGF, the *collective construction of a discourse* on global communications, where both institutional and non-institutional actors are involved, comes to summarize the whole substance of multiactor political connections. Thus, symbolic relations between themes and perceptions are sustained and, in turn, influence the very formation of multiactor political connections, whether these are collaborative or conflictual. Our analytical approach is meant to uncover what the current discourse on IG is made of (i.e., how different actors frame IG) but also how different understandings can influence the construction of political relations in the domain (i.e., the relational consequences of frames). Finally, but not less importantly, we understand connections as *historical legacies*. There is a *fil rouge* between contemporary discussions on IG and past international debates in which information and communication issues were revealed to be highly controversial topics. However, because this legacy is often over-looked, there is an overall underestimation of the effect that past experiences have on current conceptualizations of global issues and on actors' propensity to collaborate with other stakeholders involved in the same process.

It is through the joint examination of frames and connections that we engaged in the study of the IG case, thus adopting networks as our privileged investiga-tion tool. However, it is important to mention once more that this approach can be usefully applied to the investigation of other GCG domains besides the IG one. In fact, our joint use of frames and connections stems directly from the very intertwinement of contents and processes in all domains of the global communications field, among which IG is located in a prominent position. In this sense, the approach that we apply here to study the IG case was tailored by starting from the analysis of some key features of the GCG field that are isomorphically reproduced in the IG domain (Di Maggio and Powell 1983): (a) an overall *thematic uncertainty* regarding what issues should be discussed and what possible interplays exist among them; (b) a *procedural uncertainty* regarding the roles and responsibilities of institutional and non-institutional actors in the complex political management of information and communica-tion matters; and (c) the *multiplication of spaces* where these uncertainties are tackled through the development of innovative political dynamics involving the collaboration of institutional and non-institutional actors. We think that frames and connections are the most appropriate tools for improving the study of the larger composite landscape of global communications. Hence, the study of IG illustrated in this book is just one of the possible global communications studies we could perform from this fruitful perspective.

Before proceeding, we need to provide our readers with some more tools that will be useful along the journey of this book. As we said before, to avoid getting lost along the way, our equipment should include many different resources. In the remainder of this introductory chapter we will try to provide some basics elements that will help our readers become more familiar with the GCG field, the IG domain, and our approach to investigating the forma-tion of the multiactor discourse. We will start with a short overview of the broader GCG landscape to grasp its main features, and we will briefly retrace the main steps that, in the overall research context offered by the project "Communication Rights between Local and Global," led to the selection of IG as a relevant case study. After that, we will dwell in more detail on the IG issue, introducing our readers to the topic and to the main elements of our research approach. Finally, we will articulate in more detail the goals and the structure of this book.

1. THE BROADER LANDSCAPE: THE GLOBAL COMMUNICATION GOVERNANCE FIELD AND ITS STUDY

Interest in information and communication issues has grown considerably over the last years, and GCG, understood as the sum of all political processes oriented toward the management and development of information and communication issues beyond the national level, has become an increasingly relevant field for political and research practices. Certainly, such a general attention is based on the acknowledgment that information and communication are highly strategic domains for global politics (see Braman 2006; Mueller 2004; Padovani and Pavan 2008; Singh 2002). None would doubt that, from the local to the international level, possessing information and communication possibilities translates into economical, political, and social leverage (it is enough to look at big corporations such as the Murdoch group or to those ambiguous situations where media ownership is intertwined with political power, as in Italy). Moreover, the development of ICTs is linking together globalization processes, erasing distances and allowing immediate contacts on a worldwide scale. The possibility of organizing beyond space and time is certainly having an influence on how politics are conducted (Bijker 2006). In a boundless and interconnected space it is easier to coordinate at the international and regional level, and this is true both for institutions (see, for example, the EU or the Organization for Economic Co-operation and Development [OECD]) and for non-institutional actors (see, for example, the worldwide emergence and consolidation of the "no global" movement).

Grounded in the vital role of information and communications for our contemporary societies, GCG is then rapidly consolidating within the global agenda as a complex field, that is to say, as an area of institutional life (Di Maggio and Powell 1983; Martin 2003) where crucial themes pertaining to the fundamental social process of communication are subjected to discussion, to conceptual clarification, and to steering activities. Within this area, interactional patterns between institutional and non-institutional actors are tried out thus constituting "epitomes" of governance arrangements (Singh 2002), which contribute to the consolidation of the field itself but also support the reform of supranational and local politics toward enhanced democracy and open participation. Therefore, not only is the governance of global communications relevant in itself, but the way in which it is being structured also affects the very way in which politics are changing in a context that is permeated by globalization processes.

However, citizens (and institutions) are seldom aware of the existence of specific governance processes revolving around information and communication development and improvement. Even less often do they know how it is possible to contribute to these dynamics, even if open spaces for participation are set up more frequently than ever. Information and communication governance only episodically becomes part of a public discourse, and even more rarely is it analyzed in depth to uncover its substance, that is, what the issues are that fall in this relatively wide basket and the impact they have on our daily lives. The overall consequence is that not only do GCG processes remain ignored by the vast majority of citizens, Internet users, single individuals, and organizations, but the innovative features and critical points that emerge from practices developed in this area are rarely put under the spotlight and analyzed systematically. Even more obscure are the historical roots of current dynamics, not to mention the consequences on international relations from information and communication debates generated in the past and reverberating in the present. Thus the current contents of communications governance are ignored and the historical legacies, burdens, frustrations, and successes remain buried under layers of public inattention.

In general, overlooking the political side of information and communication issues can generate a dangerous bias. Indeed, we often enjoy and exploit the interconnectedness of the Internet and become informed and empowered through knowledge derived from all kinds of analogical and digital media, and yet we often forget (or ignore) that there is a strong political element that impacts the degree to which we are empowered by information and communication resources. Many policy-making dynamics, agenda-setting processes, brainstorming gatherings, consultations, governance experiments, and, also, collective action initiatives are currently occurring around media and ICTs conceived as power resources. Not only do all these dynamics involve a specific emphasis on non-material knowledge resources as tools for social and political leverage, but they also point directly to the constraints and the margins of action we are allowed within this globalized world. However, the grip of GCG issues on public discourses is still limited. In the end, there is not a broad acknowledgment of the fact that the governance of information and communication is a crucial dimension in the information society because it sets the prerequisites for our living informed and interconnected.

One of the reasons this happens is that GCG remains a challenging field for research. Attempts that have been made to systematically analyze its substance and structure as well as to communicate its relevance to the broader public have succeeded only partially. It is definitely true that interest in this area has increased over the last years. However, in study after study, we find quite some heterogeneity in the identification of the object of study: global media policy (Raboy 2002), information policies (Braman 2006), the government of electronic networks (Drake and Wilson III 2008), the management of Internet critical resources (Mueller 2002). This variety stems from a general tendency to adopt rather exclusive perspectives on this field, looking at specific topics and equating the overall complexity of the landscape with some of its aspects.

In the same way, research activities have concentrated either on deepening our knowledge on some topics or, alternatively, on the structure of peculiar processes, whereas the intertwinement between contents (old and new issues enmeshing, contaminating, cross-fertilizing) and processes (old and new modes of organizing and coordinating actions) has seldom been a matter for theoretical and empirical investigation.

Instead, in the governance of communication processes, information and communications are not only *the means* through which processes are carried on but also *the very object of political confrontation*. Research activities have to account for more than this peculiar two-tier field structure. As Raboy and Padovani (2010) point out,[1] they should also deal with two other crucial features: on the one hand, *multiplicity*, which stands for the heterogeneity of the *issues, actors, venues, and processes* that are included in the area and the *connections* among those elements; and, on the other, *convergence*, which stands for the general trend pushing different issues and modes of organization to enmesh with one another at different levels, from the local to the global. So far, no clear methodological pathways have been designed to face the challenges posed to the study of this area by these two features, neither from a conceptual point of view nor in relation to a more empirically oriented perspective. The overall consequence is that "this field of study appears to be under-theorized and open to controversial interpretations regarding the main processes and actors involved as well as the approaches and methods through which research is being conducted" (Raboy and Padovani 2010:152).

In gathering so many aspects and problematic elements, the study of the GCG is necessarily multidisciplinary and requires multifaceted approaches that, even in the rare cases when they are designed, have seldom been translated into practical research programs. As we will show below, the project "Communication Rights between Local and Global: Movement Networks and Transformation of Governance" aimed at filling in this void, thus building the overall conceptual and empirical research framework that is the very backbone of this book.

2. "COMMUNICATION RIGHTS BETWEEN LOCAL AND GLOBAL": THE ORIGINAL CONTEXT OF RESEARCH

"Communication Rights between Local and Global: Movement Networks and Transformations of Governance" was a project aimed at investigating mobilization efforts emerging locally and transnationally in relation to information and communication issues and at the crossroads between two global trends: the capillary diffusion of ICTs and the processes that have forced state actors to rethink their political role when other key players emerged on the scene. During the project, the main entry point for reflection was the absolute relevance of non-material resources, that is, information and communication, as the very matter of contention and upon which mobilization efforts have coalesced both in the form of collective action carried out by non-institutional actors and also in the form of multiactor dynamics where actors of different nature confront their perspectives on complex matters of common concern.

Our research started from the conclusion of the World Summit on the Information Society (WSIS, 2003–2005), an innovative institutional gathering convened by the United Nations under the leadership of the International Telecommunication Union (ITU). As we will explain in more detail in Chapter 1, the WSIS provided a unique international stage where information and communications issues were debated, but it was not the first event of this sort. Already in the 1970s, states met under the umbrella of the United Nations Educational, Scientific and Cultural Organization (UNESCO) to discuss the possibility of reforming global communications toward a New World Information and Communication Order (NWICO). It was on that occasion that a massive discussion on global communications started and that information and communication issues emerged as highly contested and full of political implications. After this first phase of discussion, a sort of global inattention fell like a curtain on this area until the WSIS process kicked off.

The Summit characterized itself also by its unusual format, which guaranteed access to official discussions to states and to a multiplicity of non-institutional entities coming both from the business sector and from the vast universe of the so-called civil society. In this sense, it offered a space for gathering in which actors could confront different political agendas, and it provided one of the most evident experimentations with multistakeholderism (at least in principle). Public interest and civil society entities stepped onto this institutional stage with the clear aim of assessing a completely different and innovative perspective on global communications than the mainstream vision of (some) governments and business representatives. With reference to these internal organizational and identity-construction processes, reflections on the possible emergence of a transnational mobilization on "communication rights" (i.e., on information and communication issues understood beyond their technical aspects and at the service of human improvement) began to spread within and beyond the Summit spaces (Calabrese 2004; Hintz 2007, 2009; Milan 2006; Ó Siochrú 2004a; Padovani 2005b; Padovani and Pavan 2009a). Still, when the WSIS ended there were more questions than answers. Above all, it was not clear what kind of collective effort we were witnessing: was it simply a response to government and business on the occasion of the Summit, or was it an indicator of broader mobilization efforts originating somewhere else? What were the common values and ideas joining together all the heterogeneous civil society groups? More importantly, how was this mobilization going to continue, if at all, beyond the conclusion of the WSIS process?

Thus, in the context of the WSIS, the IG thematic domain stood out among others. In general, IG can be thought of as including all those matters concerning the management and the development of the Internet, not only from a technical point of view but also in relation to the vast rainbow of uses of this technology in the social, political, economic, and legal domains. As the very backbone of the contemporary globalized world, the Internet and its management became the hottest topics within the WSIS. Discussions on IG became so conflictual and tense that the Summit space could not restore a balance (see Chapter 3). As a consequence, a specific institutional process, the IGF, was officially set up. The IGF was structured according to the same multistakeholder logic that characterized the WSIS because it was conceived as a space for fostering confrontation between governments, the private sector, and the large civil society constituency. However, unlike what happened before, it had no formal commitment to the production of binding results of any type.

Following up these major developments in the GCG field, our research project wondered if and how the mobilizations that had emerged in the WSIS context had somehow "migrated" into the more specific IG domain and, if so, what type of modification they had experienced (if any). However, in pursuing this effort we were immediately confronted with two main critical elements. In the first place, as soon as the IGF process took off, it appeared evident that what was going on there could not be adequately understood through an exclusive look at the mobilization of non-institutional actors. Indeed, the presence of states and intergovernmental actors and their commitment to the IGF process were definitely meaningful (the IGF being a UN initiative, in the end). More importantly, all participants, irrespective of their status (individuals, representatives of institutional actors, private companies, or civil society groups), were actually engaging in multiactor conversations. In this context, internal organizational processes (like those pursued by the civil society sector during the WSIS) appeared less relevant than the actual multiactor dynamic. The open feature of the Forum and the absence of any commitment to the production of binding results (together with lessons learned during the Summit) facilitated cross-sector interactions and allowed the participants to focus on how to resolve the IG *twofold uncertainty*: on the one hand, on *contents*, that is, defining more clearly the thematic boundaries of the IG domain; and, on the other, on *processes*, that is, clarifying how groups of actors of different kinds (public interest governmental organizations; private interest companies; public interest organizations, associations, and informal platforms; and the technical community) could jointly contribute to shape and implement this common IG vision, and with what roles and what responsibilities within a globalized political context (Alger 1997).

While acknowledging a need for a change in perspective, we considered it necessary to overcome the predominant analytical approach developed during the WSIS process, in which multistakeholderism was evaluated only in terms of the presence or absence of actors in or from the political process. When looking at who was included and who was left out of the discussion, existing analyses had left behind any concern for the type of interaction that had matured between and within the three sectors. Overall, the study of multistakeholderism had become reduced to the counting of different positions, and the explanatory potential of the concept ended up being too limited to lead us to a genuine understanding of how multiactor dynamics were contributing to the consolidation of the IG domain. We were then in need of a theoretical and empirical framework of analysis that could account for a significant modification within the political opportunity structure in the GCG field (Koopmans 2004a, 2004b; Kriesi 2004; Meyer 2004), which implied a maximization of the two main features of *multiplicity* and *convergence* of actors, issues, processes, and modes of organization, and that could help us understand the structure and the consequences of multiactor dynamics in the IG domain. To develop such a framework for analyzing the interplay of contents and processes, we started from a detailed analysis of the IG domain to discover how the very peculiar GCG features (i.e., uncertainty regarding thematic boundaries and political procedures) were translated within this specific area.

3. THE TWOFOLD UNCERTAINTY OF INTERNET GOVERNANCE: A CLOSER LOOK

What are the features of IG that make it such a clear case study of how the governance of global communications is being structured? For sure, IG managed to impose itself within the global agenda and to reach a particular level of public attention, as proved by the establishment of a specific UN site for its development (i.e., the IGF). This institutional recognition is perfectly consistent with the unquestionable relevance of the Internet as the main communication tool within our societies (Castells 1996; Haythornthwaite and Wellman 2002; Katz and Rice 2002). And yet, although it is now more popular than ever, IG is still a subject that draws the interest of only a few minorities. Quite paradoxically, IG is a global issue that touches upon the backbone of contemporary communications, societies, and economies but that is also affected by a "status of a minority," because of which it sounds more like food for geeks than food for politics. This paradox certainly goes back to the IG's inherent technicality: after all, IG deals with the management of the Internet and requires some knowledge about its functioning mechanisms. However, as shown by academic reflection on the progressive transformation of nuclear power from a hi-tech matter into a social concern (e.g., Gamson 1992; Gamson and Modigliani 1989; Ungar 1992), the technicality of scientific discourse has never prevented either the development of a critical public discourse on issues that were perceived as socially relevant or the realization of mobilization initiatives of various natures around them.

In the IG case, the translation of a technical discourse into a social concern and the realization of collective initiatives are still in their preliminary stages, and this, in turn, can explain the current "minority status" of this complex issue. However, the ground for an enlargement both of the discourse and of the mobilizations in this domain is definitely fertile. So far, IG has been a technical concern for only a minority of interested parties, and, therefore, the participation in governance processes strictly depended upon the possession of technical expertise and competence. Nonetheless, the seamless adjustment of global communications on the Internet infrastructure, its pervasiveness in all domains of contemporary life, and the multiplication of open and participatory processes where this richness has progressively been transformed into a collective, multiactor discourse have implied an irreversible extension of *the IG concept*. Consistently, the expertise that is needed and required for managing the Internet cannot be considered any longer in exclusively technical terms but also includes business and legal skills, competencies in defending human rights, social communication awareness, and many other types of nontechnical attitudinal resources that are necessary to the functioning of the Internet system.

It is exactly this process of multiactor construction of the IG social problem (Hilgartner and Bosk 1988) that constitutes one of the most interesting features of the recent political dynamics that have developed in this domain. Such a collective framing effort entails the shared resolution of a twofold uncertainty pertaining both to IG contents (i.e., uncertainty over thematic boundaries) and to the IG processes through which contents are translated into effective and stable governance practices (i.e., procedural uncertainty). Multiactor political dynamics are currently being developed, and more participatory political arrangements are being tried (as in the IGF case) to facilitate the resolution of the IG problem. The convergence and the intertwinement of the most disparate aspects into the same domain, together with the necessity of combining diverse competencies, call for a larger participation in political dynamics and, in the end, produce the need for a systematic change in traditional state-centered governance mechanisms. However, we have not yet reached the point at which the thematic boundaries and the procedural rules of the IG domain are all settled. If we want to systematically uncover the processes through which these two levels of uncertainty are being dealt with and the overall value of the experiments carried on in the IG domain for the reform of global politics, then it is necessary to define more precisely what the IG twofold uncertainty consists of or, in other words, to focus on the substance of the IG social problem in order to isolate its main components.

Indeed, we use the Internet to achieve different purposes, and we have different competencies and degrees of confidence in performing our activities online. Despite this immense heterogeneity, we are all familiar with the idea of "the Internet": we might not understand in depth its technical structure and mechanisms, but, in one way or another, we all know that it connects computers and people all over the world. However, only a few of us also have an opinion on the meaning of the expression *Internet governance*, while most of us simply ignore that something like that exists. This happens because we tend to consider the Internet as a matter of fact, as a network of connected computers that is "out there" and leans on some "not-better-specified" mechanisms that we do not really need to know in order to enjoy the potential of this tool. In other words, we reduce the Internet to the services that we use (our e-mail, chat, social networking sites, blogging, etc.), and we rarely wonder about what is "behind them" or, in other words, about the (infra)structure that is allowing their very existence. In any case, even among those who are aware that something labeled "Internet governance" exists, understandings and perceptions tend to differ and, to some extent, diverge. Furthermore, because we stand in an epoch where traditional decision-making mechanisms are being put into question and where multiple solutions to the retrenchment of the role of nation-states are being pushed forward, there is not one idea of governance but, rather, a "sum of crazy-quilt patterns among unalike, dispersed, overlapping and contradictory collectivities seeking to advance their goals" (Rosenau 2002:285). Even when someone argues that "everything is changing," we still do not know where the change is leading us or what it consists of. In such a context, *Internet governance* becomes more than the simple sum of two words: because of the variability of meanings that *Internet* and *governance* assume for all of us, it is impossible to define univocally what falls under this label and, therefore, to identify the thematic and procedural substance of one of the most relevant policy domains of our era.

Let's look at IG contents first. When the Internet turned into a mass phenomenon almost 15 years ago, IG broke out of the laboratory doors and progressively became a matter of interest for many others besides technicians (Hofmann 2006; Kleinwächter 2004, 2007). However, the Internet evolves with a velocity that previous technologies had never experienced, especially in terms of the uses and services built to exploit its interconnection potential. Because the contents associated with IG derive primarily from actors' understanding and uses of the Internet itself, the variability of perceptions, views, and perspectives converging on this policy domain is extraordinarily high and generates what we can call _thematic uncertainty_. On the one hand, this thematic uncertainty constitutes a major hindrance to the immediate understanding of the whole range of issues that fall under the IG umbrella and makes it more difficult to communicate its importance to the broad public of nontechies. On the other hand, though, this level of uncertainty is also functional, as it allows a certain flexibility and the fluid adaptation of the IG agenda to changes brought by Internet evolution (Drake 2004; MacLean 2004a).

So far, the boundaries of the IG domain have been set according to different perspectives. Starting from Bijker's definition of technology as a multilayered concept (2006:682) made up of _artifacts_ (physical objects such as computers and infrastructures), _human activities_ (designing, making, and handling machines and networks), and _knowledge_ (what people know about and do with technologies), it is possible to identify two main points of view on IG, each of which emphasizes different aspects of the Internet technology. A first, more restricted perspective sets IG boundaries around artifacts and human activities. One example in this sense is provided by Kleinwächter, for whom "the term 'internet governance,' while undefined, rather vague and partly confusing, stands mainly for the global technical management of the core resources of the internet: domain names, IP addresses, internet protocols and the root server system" (2004:233). A second, more holistic perspective encompasses all three aspects of Internet technology, thus emphasizing its social and human components. One example in this regard is provided by Drake, who defines IG as "the collective rules, procedures and related programs intended to shape social actors' expectations, practices, and interactions concerning Internet infrastructure, transactions and content" (2004:125).

Existing academic and political reflections have built both upon the more technical interpretation of IG and upon its more socially encompassing idea, thus, from time to time, highlighting different aspects of this domain and deriving from its study different political considerations. However, such an internal "inconsistency" is the product of the overall trend mentioned above in relation to the study of the GCG field: the preference for the technical or for the broader aspects derives from the adoption of an exclusive focus either on a particular theme or on a specific political process whose agenda is predefined and reflects participants' priorities and power levels. Instead, within a context like the IGF one, there is the unprecedented possibility of collectively reshaping the IG agenda thanks to the elimination of formal barriers to participation. In the IGF, the technical and the social perspectives on IG are given the opportunity to meet (or, more likely, to clash) and the thematic boundaries of the domain are redefined collectively through multiactor dynamics.

Consequently, occasions like the IGF offer the unprecedented chance to rediscuss IG contents and to analyze almost in real time how the thematization of the overall domain is collectively undertaken. Indeed, the IG thematic uncertainty is tackled though a collective and participatory effort to define and specify the IG meaning, thus building on contributions coming from institutional and non-institutional stakeholders. More than providing a simple definition of contents, though, setting the thematic boundaries of the IG domain means providing political actors with some fixed points of reference so that they can determine what stakes are at play, whether they feel involved, and what type of mobilization effort is required to guarantee that their interests are preserved (Abbot 1995; Melucci 1988, 1996; McAdam 1982, 2003). In other words, delimiting the IG domain from a thematic point of view is connected to the selection of those actors who are allowed to participate in those political activities that, in the end, will determine how and how much we can be informed in the global era.

In this sense, multiactor dynamics in the IG area are aimed also at making sense out of the various meanings associated with the word *governance* (Kurbalija 2008; Pattberg 2006; Rhodes 1996). In general, governance processes need to catch up with modifications of their very object of interest. When it comes to the governance of the Internet, this adaptation process must happen very quickly because of the velocity with which the tool and its uses evolve (Hofman 2006). Hence, a further level of uncertainty relates to the actual processes that are sustaining the discussion on IG contents, thus shaping the very organizational and regulatory milieu in which IG translates into practice. IG is not only a complex set of issues but also a system made up of organizations and practical decisions ranging from the assignment of domain names to the assurance of interoperability in an always growing system, from the defense of freedom of expression online to the protection of children from dangerous and offensive contents. For each of these themes there is a multiplicity of actors that claim to have an interest at stake and that, therefore, want to play a role in the discussion and the regulation process. In such a context, a traditional regulatory system based on the actions of national governments could never be adopted as the principal management modality because of "the international and decentralized nature of the Internet" (Baird and Verhulst 2004:59) and of the overwhelming thematic variety that characterizes this domain.

Nonetheless, the fact that traditional policy-making mechanisms are not suitable for facing the challenges that characterize this area does not imply that rules are not needed. It means simply that different ways of producing rules and coordinating the different parts of the Internet system must be found. However, "so far, attempts to establish a stable governance arrangement have failed. Over the last fifteen years there have only been phases in which the transformation of governance structures has slowed down" (Hofmann 2006:2). The practical arrangements that have been proposed do not seem to have reached the necessary level of flexibility that the governance of the

Internet system requires. The capillary diffusion of this tool into houses and the progressive transformation of domain names from technical elements into objects for speculation limited the effectiveness of purely technical management solutions (Botzem and Hofmann 2008; Hofmann 2006). Also, attempts to institutionalize self-governance mechanisms that avoid any interference from governmental actors were unsustainable because of the need for international synergies and the indisputable necessity to protect users according to national legal systems. At present, governmental and non-governmental entities are partners in the effort to dispel this *procedural uncertainty*: they must find a mutually satisfactory way to harmonize their activities in the IG area and thus coordinate technical development and regulation with societal and economical needs. This effort requires continuous experimentation with interaction and collaboration practices in order to reach an overall arrangement that, on the one hand, is flexible and efficient enough to satisfy different actors' needs and instances and, on the other, can fluidly adapt to the evolution of IG contents. The challenge that is at stake here is not an easy one: it consists in finding an agreed-upon pattern for moving toward the unquestioned repetition of the "sequences of standardized interaction" (Jepperson 2000:198) that govern the political behaviors between IG agents in the context of a shared symbolic universe.

There is a further element of complexity that adds to the thematic and procedural uncertainties, namely, the evident interplay that exists between these two dimensions. The looser the issue boundaries and the more heterogeneous the constellations of actors within the IG area, the more encompassing and inclusive the processes will have to be to coordinate their actions. Thus, the more different stakeholders that find their place in the game, the more the collective construction of IG meaning will depend on continuous renegotiation over contents. To put it simply, ongoing consolidation of IG at the global level depends on and varies according to the continuous interaction between content and process that are both socially and dynamically determined.

4. THE REFORM OF WORLD POLITICS AND THE CONSTRUCTION OF AN INTERNET GOVERNANCE DISCOURSE

The provision of an open space for dialogue like the IGF is only the most recent evidence of the widespread adoption of the multistakeholder approach to the conduct of politics as a supplement to (not a substitute for) traditional state-centered policy-making mechanisms. Indeed, in an overall context where "society is no longer exclusively controlled by a central intelligence (e.g. the State); rather controlling devices are dispersed and intelligence is distributed among a multiplicity of action (or 'processing') units" (Kenis and Schneider

1991:26), promoting processes whereby all necessary *intelligences* are working together to enhance the efficiency of political mechanisms becomes a necessary attitude. As we will see in more detail in Chapter 1, we are standing in an epoch of overall political uncertainty (Robertson 1992) caused, in the first place, by the fact that states have to face multiple and complex issues that extend beyond the boundaries of their sovereignty and, more importantly, that require an incredibly large amount of competency to be managed adequately. This does not mean that states have lost their functions: institutions continue to be the sole agents in charge of producing policies. What changes is that they can no longer perform their functions "behind closed doors" but, rather, find themselves forced to act within a very crowded environment, populated by a multiplicity of non-institutional actors who possess the required knowledge and the expertise for managing complex and dynamic global issues (Kooiman 2003). How to translate the necessity for multiactor collaboration into efficient governance arrangements remains an open question.

This is particularly true in the case of information and communication matters, where technical and social aspects are both relevant and so interwoven that, when it comes to their regulation, governments have to coordinate a plurality of interests, knowledges, agendas, and priorities but often are not equipped with the necessary competencies to do so. In the Internet case we have the extreme situation in which governments were also the last actors to be involved in the management of a system that had self-managed itself for years (Hofmann 2006; Kleinwächter 2004, 2007). The underlying question of how we shift from "government to governance" (Rhodes 1996) in the IG domain becomes, in general, a question about how we can effectively relate traditional steering activities, for which states are responsible, with broader coordination tasks that go back to the very origin of the Net itself and that combine the multiple perspectives and needs of all institutional and non-institutional Internet users. What is the role of different actors' categories in the governance dynamics of the Internet? How to (re)conciliate perceptions, positions, and political interests?

Current discussions on IG and the multiactor dynamics that are taking place within spaces like the IGF provide a useful case study for understanding not only how this specific domain is evolving but, more broadly, how world politics are changing toward more open and participatory arrangements. However, to achieve this better comprehension, it is crucial to approach the study of the IG domain in a systematic way, one that is able to open the door for a critical reflection on how arrangements in global politics are evolving in general and, at the same time, to critically account for the actual changes occurring in this area.

Consistent with this need, this work will explore from different angles how IG is being consolidated as a relevant political domain through the development of transnational multiactor dynamics that are aimed at reducing its twofold uncertainty. In our exploration we do not look at traditional policy-making processes in this domain or at the actual regulation of specific topics within it (e.g., Internet critical resources management, privacy, etc.). Rather, we investigate the IG domain from an alternative entry point: the progressive construction of an *IG discourse* that is fostered by the realization of the IGF, that is, a privileged space where the thematic boundaries of the IG domain are being defined together with interactional patterns between actors that have a stake within it. In the context of our study, discourse must be understood as "metaphorically extended from its original roots in interpersonal conversation to the social dialogue which takes place through and across societal institutions, among individuals as well as groups and . . . political institutions themselves" (Donati 1992:138). For the IGF does not have a steering function; it is only meant to provide a "forum for multi-stakeholder policy dialogue" (WSIS 2005:art.72), the *construction of the IG discourse comes to summarize the essence of political activity* because it entails the production of a set of "concepts, categories, ideas, that provide its adherents with a framework for making sense of situations, embodying judgments and fostering capabilities" (Dryzek 2005:1). Looking at the construction of the IG discourse, we then look also at how the IG social problem is collectively constructed and, ultimately, at how institutional and non-institutional entities end up working side by side on a set of issues that are of common concern.

In investigating multiactor discursive dynamics aimed at *framing* IG (Goffman 1974; Snow et al. 1986; Snow and Benford 1988), this work does not deny that within the IG domain conventional steering processes are also taking place and that, in this sense, IG can be studied as a policy domain in the most traditional way (Knoke et al. 1996).[2] However, the establishment of innovative political spaces where actors and issues converge to collectively shape a common discourse and where there is an inevitable tension between the conservation of the status quo and innovation in actual governance arrangements does actually have an impact also on more traditional policy-making activities. Indeed, as we mentioned above, this work starts from the belief that transformations in the ways of framing IG are full of implications for the very functioning mechanisms and for the future developments of the IG domain itself. In other words, the multiactor construction of a new IG discourse is crucial because it provides the base for the emergence of norms, understood as "standards of appropriate behavior for actors with a given identity" (Finnemore and Sikkink 1998:891), that in turn will ground the future regulation of the Internet system as they will determine the roles and the responsibilities of the different stakeholders and the contents of the political agenda. Therefore, from a research point of view, the examination of dynamics fostered by the IGF context is particularly indicative in relation to the process of overcoming state-centered mechanisms because it points directly to the question of how and when behaviors become "appropriate," that is, when norms consolidate and ground regulation activities in a certain area (Khagram, Riker, and Sikkink 2002; see also Faist 2004 and Della Porta et al. 2006).

It should be remembered that norms, defined as above in terms of appropriate behaviors, do not constitute per se "hard" forms of power (Keohane and Nye 1977), but they nonetheless are far from being weak, as "the power to shape the agenda, or to shape the very manner in which issues are perceived and debated, can be a substantial exercise of power" (Sikkink 2002:303–304). In this sense, if we adopt Melucci's definition of political relations as those "which are activated in order to reduce uncertainty and mediate among conflicting interests through decisions" (1996: 211), then it becomes evident that multiactor interactions developed within the IGF space are *inherently political*. In this specific case, decisions will pertain to IG contents as well as to the roles and responsibilities of governments, business entities, and civil society groups in the IG domain. Thus, these decisions will not be codified through binding provisions but crystallized in the institutionalization of political behaviors that will emerge from experiments on the ground in the formation of an IG discourse. Once "appropriate" behaviors stabilize over time, they will support mutual recognition and collaboration efforts in *governing* (i.e., regulating) the Internet. Moreover, the fact that these multiactor interactions entail a predominant symbolic component and are devoid of physically violent confrontation by no means implies that their political relevance is minimal or that conflict between the actors involved is less severe (Melucci 1996). Indeed, the "creation and implementation of institutional agreements are full of conflicts, contradictions and ambiguities" (Meyer and Rowan 2000:46). Therefore, there is no reason to think that the construction of a discourse over IG will be plain and peaceful only because no formal policy-making activity is foreseen in the IGF context or because there is no violence in the repertoires adopted by participants. The important question then is not whether the construction of the IG discourse is conflictual but, rather, what form that conflict is taking in a symbolic challenge like this one and how mediation to overcome disagreement will actually happen (if at all).

How do we translate a theoretical focus on the formation of discourse on IG (and global information and communication issues in general) into a framework for analysis? How do we account, simultaneously, for the evolution of contents and processes in the IG domain as well as for their intertwinement within multiactor practices? To achieve these goals, we mainly lean on two concepts: *frames* and *connections*. In our study, we use the concept of frames to uncover the political substance of IG discourse, while we understand connections, at the same time, as interpersonal, multiactor ties, symbolic ties, and historical legacies with past debates on information and communication issues. Frames and connections are, in our approach, what ground the current development of a multiactor IG discourse, and, more generally, we argue that they are two conceptual tools that can be fruitfully applied to the study of GCG domains.

We join frames and connections together through the adoption of a relational approach based on the systematic use of networks as analytic and empirical tools for investigation. Following Mische (2003), we understand networks as the locus where the IG discourse is created. Consistently, networks are employed here both as a powerful image to depict the complexity of the phenomena under examination (Kenis and Schneider 1991) and as an analytic tool that allows us to concentrate simultaneously on the multiplicity of actors and themes in the field and on the interactions they engage in. The central role played by networks in our study is a consequence of their nature as flexible organizational modes: indeed, they arise in response to the overall uncertainty characterizing the global society (Powell 1990; Jones, Hesterly, and Borgatti 1997) and allow us to manage the different types of interdependencies that are established between actors (Hockings 2006). This characteristic of the network organizational mode is totally suitable for depicting the convergence of actors within the IGF space in order to resolve the IG twofold uncertainty and to exchange the knowledge and competence that is required to manage the different aspects of this domain.

Cognitive and knowledge interdependency between the actors involved in the creation of an IG discourse (in general, in all global political processes) translates, in the first place, into the construction of communication relationships among those who do possess the necessary resources to deal with specific problems. Hence, networks that arise in the global landscape can be considered in terms of *communication networks* or "patterns of contact that are created by the flow of messages among communicators through time and space" in the attempt to stabilize structures of interaction out of the chaos provided by the globalized contexts (Monge and Contractor 2003:3). In relation to the production of a discourse in the IG domain, communication networks are conceived as the loci where IG contents and processes come together for the very production of norms: in this sense, they are shaping, and, at the same time, shaped by cultural and communicative interaction between actors themselves (Mische 2003:258). Looking at the formation of an IG discourse within and through networks, we argue that it is possible to go beyond the descriptive feature of research efforts that have been pursued in the past in relation to the study of the GCG field by leaning on the concept of multistakeholderism as the preferred perspective (Padovani and Pavan 2011). Thus, networks allow us to grasp to a deeper extent how multiactor dynamics are being structured and how this impacts the configuration of the IG domain in the global political landscape.

Furthermore, the relational perspective we adopt here allows us to explore the formation of the IG discourse across a traditional boundary that is characterizing the study of politics in general: that between the virtual or online world and the real or offline world. Interestingly enough, despite the emphasis put on remote and online participation within the IGF context (de la Chapelle 2010), IG dynamics so far have been examined mainly within the offline space, whereas online interactions among participants to the Forum have been often overlooked. Because of the very nature of the Internet communication tool, the virtual dimension must become part of the analytical picture we are trying to take. The online space provides a further environment for increasing participation in political processes, one in which the traditional resource constraints that hamper a physical presence exert a less dramatic impact (Padovani and Pavan 2008). Indeed, once appropriate platforms are set up, technical requirements are met, and basic skills are provided, potentially everyone is able to contribute to a pluralistic online conversation. Consistent with this perspective, online interactions will then be included within our analytical effort as they contribute to framing IG issues and shaping actors' interests (Padovani and Pavan 2007). The online and offline dimensions constitute for us different *territories* upon which the IG domain is consolidating: they are complementary, but they host different actors and relational dynamics (Rogers 2006). Furthermore, there is no hierarchy between the two levels, even though it is easier to think about the online space as a projection of the offline one rather than the other way around. In this sense, they will be explored on their own as well as in their intertwinement: results obtained by these two explorations will be read jointly to derive a more complete picture of the IG landscape.

5. AIM AND OVERVIEW OF THIS BOOK

Existing reconstructions of IG dynamics do take into account the processes of norm production, the allotment of different tasks to different organizations or individual actors within the complex IG domain, and, sometimes, even the ties existing between IG contents and processes. However, these elements are seldom considered all together. In fact, there are several accurate attempts to map the different functions entailed by IG mechanisms as well the actors appointed to exercise specific functions (see MacLean 2004b; Mathiason et al. 2004). Despite the inherent value of these contributions, existing reconstructions have all started from an a priori definition of IG that was somehow imposed because IG can encompass virtually anything and everything that involves communication and information (Mathiason et al. 2004). In these cases, limiting the observation to some crucial aspects of this vast political *mare magnum* was a necessary preliminary step. In this work, instead, we adopt a bottom-up approach: we start from practices that develop in the field and that consist of the joint collaboration of institutional and non-institutional actors and from these we infer where the boundaries of the IG domain are being set. We will do so by exploring how frames and connections are being developed within the IGF space and in the online and the offline spaces. However, as we mentioned above, we think that the aim of this book is broader than the sole illustration of results obtained from empirical analysis of the IG case grounded on network tools. More generally, this work is meant to provide concrete evidence of how to account for multiplicity, interconnectedness, and convergence of actors and themes in the current governance of global communications.

Some of our readers might think that this book is all about the methods we employed in our investigation: how we operationalized variables, identified nodes, traced ties among them, and so forth. In fact, this is neither a book exclusively *on* IG nor a book *on* network analysis. Rather, it is a book that *looks at* IG *through networks,* but it does not adopt an exclusive focus on either of these two elements. Even if specific methodological choices and results coming from our study of networks in the IG case are at the core of this work, mastering our knowledge of this specific domain through an innovative research approach is not the sole task we are pursuing here. Beside (and beyond) that, this study would like to provide concrete and systematic evidence of the fact that information and communication issues like IG are pivotal parts of the global agenda and that, although they are often labeled in non-obvious ways, they subsume many other concerns that are more familiar to us and that we consider central in our daily lives (for example, universal access to the Internet, the digital divides, freedom of expression, privacy, and security, just to mention a few). Given their relevance, information and communication issues need to be studied systematically in order to uncover what their political agenda is made of and who the actors are that are shaping it. In other words, this study wants to provide evidence of the fact that global communications and IG are not only food for geeks. They are also (good) food for politics and for research in social science.

Starting from these premises, we can articulate three different objectives of this book: (a) raising awareness of the centrality of information and communication issues as complex global policy fields, thus providing an overview of their main development over time; (b) highlighting the challenges connected to the systematic study of GCG, thus offering an original and innovative approach for its research; and (c) investigating a relevant case study, that of IG, to shed new light on it but also to understand in more detail how the governance of global communications is currently being structured. To conduct our investigation at these three levels, we adopt in this book (as we did in the project) a bottom-up perspective, one that is rooted in the idea that the governance of global communications, that is, the sum of all political processes oriented toward the management and development of information and communication issues beyond the national level, is formed by multiple domains that crosscut, intertwine, and influence one another. Domains constitute the starting point for our research activities because specific policy processes, dynamics, challenges, and claims are actually happening with specific reference to these smaller (but not less complex) areas. We believe that it is starting from the empirical study of domains that the overall GCG landscape can be mapped in a journey that is certainly long and demanding but that is important to take.

In this sense, we consider IG one crucial domain in the GCG field, and, as we have seen, there are many reasons why it provides an interesting case study for inquiring into this global policy field but also into the changes of supranational political mechanisms centered on the nation-state. Still, since our perspective implies that IG per se does not exhaust the whole GCG field, the concepts and the analytical approach we propose here are also used for the exploration of other domains and areas of global communications. The real added value of this book, then, is not exclusively provided by the results obtained from the investigation of our chosen case study. In fact, a further value comes directly from the research framework we elaborated to systematically study a broad field that, although it has received increased attention, remains in need of sound theoretical conceptualizations and empirical investigations. In other words, as suggested by the title, the added value of this work derives not only from the network study of IGF dynamics but, more generally, from the joint use of frames and connections to investigate complex domains such the IG one as part of the broader governance of the global communication landscape.

In sum, this is a book that builds, at the same time, on frames, networks, and the centrality of IG to think about the dynamics of interaction between institutional and non-institutional actors that are established when issues of common concern are dealt with. The chapters that follow this Introduction reflect this overarching goal and explore the three tasks that we assigned to this book. In the first place, we said, this book is meant to raise awareness of information and communication issues as complex global policy matters that crosscut broader reflections on how global politics are changing. A necessary first step, then, is to reflect on the context of broader transformations of politics and to illustrate how the GCG field has evolved over time. This reconstruction effort, although complex, is necessary for two reasons. First, because it summarizes the state of the art of research that is developed with reference to the governance of global communications, such an overview is necessary to make sense of the framework that we propose in Chapter 2 and that we apply in the second part of the book to specifically analyze IG dynamics. Second, the story of global communications is a story of globalization, international relations, political games, and international conflicts that is not told as often as it should be and, therefore, from which we do not learn as much as we could. Chapter 1 responds to this twofold necessity by drawing the overall background to our analysis of the IG case from both a structural and a historical point of view. First we look at the intensification of globalization processes, the discussion on global governance, and the reflection on one of its possible realizations, namely, multistakeholder processes. Second, as this work examines the consolidation of the IG domain with a specific focus on official spaces for political debate on information and communication issues, Chapter 1 reviews the two main institutional occasions for discussion at the international level: the debate on the establishment of a New World Information and Communication Order and the World Summit on the Information Society. Finally, we join together the two strands of reflection on structural and historical elements of the GCG with the aim of pointing out the challenging points in the study of this field.

This leads us to the second objective of this work, that is, offering an original and innovative approach for theoretical and empirical investigation. Instead of looking separately at the contents and process of the GCG (and of IG in particular), this work joins them together in the networks of interaction where the discursive dynamics are taking place. However, since we cannot start from the assumption that all readers are familiar with the conceptual tools that ground our approach, Chapter 2 provides a twofold overview of both frames and networks (which are the operational tools we use to render the idea of connections). In the first place, it clarifies the concept of frames and outlines a working definition that serves as a starting point for a closer examination of their relevance within political dynamics. It then illustrates the main premises of a relational approach to the study of social phenomena and the general conditions in which networks tend to emerge as distinct, specific organizational modes. Furthermore, starting from the idea that contemporary politics are played out on the basis of interdependencies between actors that facilitate the construction of *communication networks* (Monge and Contractor 2003), Chapter 2 reviews four network concepts that have been elaborated in literature to depict different situations of interdependency between actors in politics: *government networks, policy networks, collective action networks,* and *governance networks.* We organize these four concepts along two dimensions—actors' diversity (i.e., homogeneous networks and heterogeneous networks) and the goal of the political process (i.e., binding outcomes versus non-binding outcomes)—so that each of them comes to represent an ideal typical situation of political interaction. Among these ideal types, we then choose *governance networks* as the most appropriate perspective for grounding the framework of analysis employed to explore the collective and multiactor construction of an IG discourse. Starting from these bases, our analytic framework considers the online and the offline dimensions as two separate, yet interrelated, discursive spaces, and, for each of them, it foresees the exploration of two types of networks that correspond to the two dimensions of uncertainty affecting the IG domain. On the one hand, our framework entails the investigation of semantic networks, which will "map similarities among individuals' interpretations" (Monge and Contractor 2003:173).[3] On the other hand, our framework entails social networks, which will be explored as they are constituted by a "finite set or sets of actors and the relation or relations defined on them" (Wasserman and Faust 1994:20).

The remainder of the book is then dedicated to the pursuit of the third objective: investigating a relevant case study, that of IG, to shed new light on it and to understand in more detail how the governance of global communications is being structured. Chapter 3 introduces in detail the specific domain under examination with the overall aim of clarifying how its thematic and procedural uncertainties evolved over time. Specific attention is paid to outlining the main characteristics of the political opportunity structure provided by the IGF and, more particularly, to Dynamic Coalitions (DCs), that is, multiactor groups that originated in the IGF context to foster the articulation of discourses in relation to specific thematic areas of the IG domain. Also, some general notes are provided on how data for this research were gathered. Finally, Chapters 4 and 5 illustrate how social and semantic networks were explored, respectively, in the online and offline spaces. Each chapter concludes with a snapshot of the two discursive spaces where results of the examination of the two different types of network will be compared.

We like to think of this book as a unique storytelling effort that moves from the history of global communication in the 1970s to the network analysis of the multiactor dynamics fostered by events such the IGF. Certainly, those readers who are more interested in the IG case and the empirical phase of our investigation can skip directly from here to the second part of the book and leave aside the historical and theoretical journey that is grounding it. Conversely, those who are not familiar with frames, networks, and the governance of global communications or with how they can be employed in the study of political processes can switch directly to Chapter 2 and leave for later the rest of the global communication story we tell here. However, as in all stories, there is always a precise reason why protagonists behave in a certain way rather than in another. We therefore think that all three parts of this book (the story, the theory, and the fieldwork) are necessary for a correct understanding of our proposed analytic framework as well as of the results of our exploration of the IG domain. We invite our readers to engage in the same journey we took and that starts from the exploration of the broader context to then reach a particular spot in the land of global communications. All together, the chapters in this book prove that different competencies, methods, and research questions can render a more fine-grained picture than the one that will result from adopting just one disciplinary perspective. Our theoretical and empirical results should then be considered as a starting point for future research activities in this field, and we hope that our experience will contribute to a genuine dialogue *on* global communications that is also *made of* global communication between researchers.

NOTES

1. Raboy and Padovani do not refer to Global Communication Governance (GCG) as their object of interest but, rather, to global media policy (GMP). However, the conceptualization of the GMP field they propose is totally consistent with our view on GCG.

2. Policy domains are defined as consolidated areas of interests where participation is restricted to consequential actors, namely, those who have a clear interest in the issues discussed (Knoke et al. 1996). Thus, the final outcomes of interaction are usually directed toward the regulation or legislative activity upon which different actors try to exert the highest influence.

3. As Monge and Contractor explain in their work (2003:186–188), the idea of a semantic network as a specific type of communication network serves the purpose of clarifying "the relationship between communication and shared understanding" (Monge and Contractor 2003:187), thus problematizing the idea that communication between actors leads to shared interpretations and understanding. Also, the authors specify that this understanding of semantic networks stems from, but also differs from, Carley's idea that concept networks are networks of concepts or "ideational kernels" and the "pairwise relations between them" (1997:81).

Chapter 1

A Background Picture

The emergence of Internet governance (IG) as a distinct domain of discussion and political confrontation has taken place over the last few years in a context of growing attention for information and communication issues on the global scale. In turn, this has been stimulated by even broader conversations on ongoing changes in political arrangements at the supranational level, in international relations, and in the modes of participation of non-institutional actors in politics. For these reasons, in order to make sense of the analytic framework that we are going to apply to our exploration of the IG case, we have to start with a close examination of the larger context in which this particular topic has consolidated. This is a complex exercise as the picture taken into consideration is composite, consisting of several intertwined structural and historical elements, each of which deserves to be analyzed and put in relation with others. In this chapter we pursue this effort by adopting a double perspective: the first, on the structural changes that have occurred within politics that have resulted in the retrenchments of traditional state-centered mechanisms; the second, on those international occasions in which information and communications issues have been officially discussed extensively.

As far as structural elements are concerned, it is now a well-known fact that Information and Communication Technologies (ICTs)—and the Internet in particular—are fundamental globalizing factors and that they have been crucial in undermining the centrality and the legitimacy of the nation-state as the sole locus of political authority. Indeed, in a context where the requests that the state has to answer are increasingly numerous and demanding, and involve the most disparate sectors (e.g., economics, health, society, technology, etc.), ICTs sustain the emergence and the consolidation of the non-institutional actors who own the necessary knowledge that governments are missing to perform their coordination and steering tasks. It is precisely in the attempt to reconcile the loss of authority of the state with the richness of political resources derived from non-institutional entities that several *governance arrangements* have been experimented with over time. However, as we will see below, *global governance* remains a contested idea characterized by a conceptual heterogeneity that hinders the systematic assessment of different solutions adopted to manage the increased complexity of the global situation. In the first part of this chapter, we will pay specific attention to one governance arrangement, namely, the so-called "multistakeholderism," understood as those forms of interaction that can occur both at the national and at the supranational level, that are aimed at managing a certain issue, and that intentionally involve a plurality of different actors with a stake in it.

Looking instead at the historical elements, in the second part of the chapter we will review two past institutional occasions for discussion hosted at the international level and fully focused on information and communication issues: the New World Information and Communication Order (NWICO) debate, carried on by ex-colonies in the 1970s; and the United Nations–sponsored World Summit on the Information Society (WSIS), which took place in two phases (Geneva 2003 and Tunis 2005). Although some key differences exist between these two institutional processes, in the first place they both contributed to progressively mark the boundaries of Global Communication Governance (GCG) as a global domain wherein diversified and complex political dynamics can develop, and, secondly, they showed the contested feature of information and communication issues as possible sites of struggle. During the NWICO debate, the nexus existing between media, communication, and world imbalances was made explicit for the first time. The overlap between tensions derived from Cold War dynamics and those referable to the very matters discussed in relation to the establishment of a NWICO made of international information flows a minefield upon which nation-states confronted one another and national interests clashed. Instead, the WSIS characterized itself as being an explicit multistakeholder experiment that joined together governments, the private sector, and civil society or public interest entities. It is exactly for its peculiarity that observers focused predominantly on the process rather than on the contents discussed.

Concluding this first chapter, we will try to join the structural and the historical elements so as to identify the main continuities and the changes in the GCG field, looking at both the contents and the international processes. Starting from continuities and changes, some key challenges to the study of the GCG field will be identified. In this way, we will sketch out some preliminary research questions to the exploration of the vast IG case, thus building a conceptual bridge toward the next chapter, where some analytical and theoretical tools grounding our analytical framework are examined.

1. THE USUAL SUSPECTS: GLOBALIZATION, (GLOBAL) GOVERNANCE, AND THE MULTISTAKEHOLDER APPROACH

When speaking about structural changes in supranational arrangements, it has become quite usual to call into question two concepts: *globalization* and *global governance*. More recently, *multistakeholderism* has become another jargon expression to identify attempts to join together institutional and non-institutional actors in multiactor dynamics to manage global concerns, especially (but not exclusively) in relation to information and communication issues. Global governance, globalization, and multistakeholdersim are, as we call them here, the "usual suspects" that are always involved in the analysis of specific cases in the global space of politics. Since, in one way or another, they always enter the picture, we think that a first necessary step on the way to the illustration of our analytic framework is to identify their main conceptual components.

1.1 Globalization Processes and the Globalization of the Political

It has been some years now since the word *globalization* entered our daily lives. We take the fact that we live in a globalized world almost as a given: we experiment with globalization every day even though we do not realize it, and we contribute to and are affected by it even though we might not be aware of our involvement. As some observers outlined, *globalization* is a word that is overemployed, first because of the pervasiveness of the phenomena but also because of its vagueness (Padovani 2001). Two inherent features of our globalized way of living are the fact that the world is progressively transforming into a "shared social space" that is shaped and allowed by communication technologies and economic forces (Held et al. 1999:1) and an overall sense of insecurity or, to use Robertson's word, "uncertainty" that is due to the diffused perception that governments are not efficient or effective in facing global challenges and that we have consequently lost our political points of reference (Robertson 1992).

For the purpose of this overview, globalization can be considered along the lines of the definition provided by Held and his coauthors, as "a process (or set of processes) which embodies a transformation in the spatial organization of social relations and transactions—assessed in terms of their extensity, intensity, velocity and impact—generating transcontinental or interregional flows and networks of activity, interaction and the exercise of power" (Held et al.

1999:16). This definition contains indeed all the elements that are necessary to define more precisely the boundaries of the "globalization idea." In the first place, it questions the fact that globalization is necessarily one process with a univocally defined end (i.e., the creation of a free and global market). It is indeed more appropriate to speak about *globalization processes* in the plural form, thus acknowledging that they actually deploy in a plurality of realms (politics, military issues, economics, culture, migration, and environmental concerns) without necessarily generating either homogeneous or unique consequences. Secondly, this definition links the analysis of globalization processes to a set of specific features that characterize them in comparison with other contemporary phenomena, such as internationalization and localization:

- The *extensity* of social, political, and economic activities;
- The *intensity* of interconnections established between different actors and activities;
- The *velocity* of exchanges, which is always increasing as a consequence of improved communication and transportation systems;
- The *impacts* of global processes and dynamics on the local dimension.[1]

Finally, this definition puts at the core of globalization processes the *networks of activity and interaction* that develop in various realms and whose characteristics (in terms of extensity, intensity, velocity, and impacts) determine the different forms of globalization we are witnessing (e.g., intense global market exchanges in the financial arenas, culture enmeshment processes, etc.).

Within such a complex scenario, some relevant transformations have definitely mutated the overall context of action for states "as their power, roles and functions are rearticulated, reconstituted and embedded at the intersection of globalizing and regionalizing networks and systems" (Held et al. 1999:440). In particular, Held and his coauthors identify five principal strands of change sustaining the consolidation of "global politics" (1999:80–81):

- The multiplication of "loci of effective political power" beyond the sole national governments and a consequent redistribution of power among a plurality of forces and agencies at the national, regional, and international levels;
- The boundless character of modern communities, which are not exclusively contained within national borders but can be grounded in shared identities and practices, as reflected, for example, in concepts such as "communities of practices" (Sassen 2004) or "enclaves" (Calhoun 1998);
- The displacement of sovereignty from its unitary, state-centered feature into a plurality of authorities and agencies;

- The complications that are brought by the rise of transnational problems (such as environmental degradation or the coordination of market flows) that can hardly be managed by calling upon a "reason of state" tied to the existence of national boundaries;
- The difficulty of marking distinctively domestic and foreign affairs given the deep enmeshment between the local and the global.

In sum, "in the political system, globalization has brought a transnationalization of political relationships" that has transformed "the international system based on the nation-state . . . into a political system composed of overlapping multi-level authorities" (Della Porta et al. 2006:43). Still, there is not one unique response to the challenges posed by globalization processes to the core role occupied in the political system by the nation-state. Several solutions for managing the complex ensemble of newly emerged actors and authorities have been pushed forward in order to face the overall uncertainty that results from global structural changes. Despite their internal heterogeneity, they share some core features that make them fall under the category of "global governance experiments."

1.2 Global Governance

The complex rearticulation of the state-centered working logic has been often depicted with the same "vagueness" that characterizes the analysis of global trends it derives from. The label *global governance* is often employed to indicate emergent ways of coordinating political action within globalized contexts, thus resulting in a catch-all formula recalled anytime we discuss arrangements adopted in the contemporary and globalized context. Since the very birth of the concept, observers have sometimes suggested that, despite the popularity of this label, its meaning and utility are to be questioned (see e.g., Finkelstein 1995). This is mainly due to the intense and diversified use of the term *governance*, which hampers the application of neat boundaries to the various emergent forms of political coordination (Kurbalija 2008; Pattberg 2006; Rhodes 1996).[2] The fuzziness of the governance concept makes it actually difficult to distinguish innovative governance realizations from already existing forms of agency. In this sense, Latham points out, for example, that it is not so clear how global governance differs from *resistance* or *contests* and that, perhaps more importantly, the "implications" of governance projects remain obscure, for example, with reference to the identification of *winners* and *losers* or in relation to the chances to question process outcomes whenever they are considered unsatisfactory (1999:49).

Certainly, the examination of the possible answers and of the practical attempts that have been proposed to manage political globalization is far from being an easy task. Moreover, the study of governance experiments has not always been pursued systematically, and this has meaningfully increased the skepticism toward this concept. One possible way out of this fuzziness is to identify the core elements that unify different concrete occurrences and that can serve as the starting points for systematic conceptualizations and empirical analysis efforts. Toward this aim, we start from Pattberg's definition, according to which governance is "first occupied with rules, organization and the conditions for order in a broader sense. Secondly, [it] stipulates the existence, to various degrees, of new processes and mechanisms of problem-solving. Finally, it describes a qualitatively new relation between public and private actors and a broadening of governing capacities, often in the form of self-organizing networks" (Pattberg 2006:5). Hence, to look more systematically at global governance realizations, we could organize our analytic efforts around three core elements:

- *Causes*: Why are actors engaged in relations with other partners in order to achieve organizational and regulating tasks?
- *Actors*: Who are the actors engaged in relations?
- *Interaction*: What kind(s) of relations are actors engaged with?

Causes: Why Are Actors Engaging in Relations?

Because contemporary policy matters are increasingly characterized by growing *diversity, dynamics,* and *complexity,* it is not realistic to assume that governments can efficiently tackle issues that constitute common concerns and touch upon the interests of multiple actors all on their own (Kooiman 2003). It should be made clear that what we are witnessing is not a complete defeat of the state but, rather, a progressive renegotiation of its traditional role that parallels the redefinition of modes for pursuing regulative and policy-making tasks. Instead of dominating political processes, governments find themselves in a situation for which they become more often *facilitators* who involve new partners from the market (i.e., big transnational corporations or powerful industries) and from civil society in activities that previously were exclusively their own (Kooiman 2003). The necessity for enlarged and collaborative experiments can be linked to a "threefold deficit" that is being experienced by political agents—especially those of an institutional nature— and that is, at the same time, a lack of *legitimacy, knowledge,* and *access* (Hockings 2006:21). The *legitimacy deficit* refers to the diffused decline of confidence in traditional mechanisms of representative democracy. The *knowledge deficit* refers to the always increasing levels of knowledge, skills, and creativity that are required to efficiently tackle global problems and that are owned exclusively by one sole sector within society. Finally, the *access deficit* refers to the enduring predominance of institutional actors inside political spaces despite the push of participatory requests linked to the legitimacy deficit. In this sense, this latter form of deficit affects non-governmental entities that claim to be part of the process, but, as with the other two deficiencies just illustrated, it results from the incapacity of states to renew decisional mechanisms so as to include non-traditional and non-institutional partners.

Hence, when we analyze concrete episodes where global governance experiments are played out, we should first and foremost look at their causes; that is, we should ask questions about the very reasons why states cannot be left alone in facing the problem under discussion: What are the different knowledges and competencies that would be required to efficiently manage the problem? What are the benefits of setting up a cooperative system instead of relying on centralized regulation mechanisms? What would be the losses if state-centered mechanisms were preferred?

Actors: Who Are the Actors Engaging in Relations?

In the globalized context there is a proliferation of actors (carrying both private and public interests) who are progressively finding their places within political processes and thus filling in the vacuum left by the shortcomings of traditional state-centered mechanisms (Della Porta et al. 2006:12–14). However, the simple acknowledgment that more heterogeneous and non-traditional political actors have become part of political dynamics does not fully answer the question about transformations of politics in globalized contexts. In fact, more than the differences determined by attributes characterizing actors in global processes, it is important to look at how these differences translate into different possibilities or constraints to concrete action. Indeed, actors' identities and characteristics are not defined once and for all: quite the opposite, they tend to change depending on the very interactional context they participate in. In his analysis of governance processes, Kooiman speaks of *sociopolitical governance* to indicate the enmeshment between public (institutional) and private (business and/or civil society) actors within new modes of organizing governing functions. Thus, he particularly stresses the paramount relevance of a focus where actors *and* the interaction they engage in are considered as the two sides of the same coin: "actors are continuously shaped by (and in) the interaction in which they relate o each other. . . . Taking a closer view, actors might themselves consist of interactions, and the boundaries from which they derive their identities are relative and often ambiguous" (2003:17). In this sense, the proliferation of actors on the political scene is much more than a mere multiplication of agents. It rather takes the form of a complex interaction system in which "actors in relation" are at the core and actors' diversity continuously increases as their identities are defined interactively and constantly renegotiated. Roles, functions, and power are then played out within interactional systems of relations established among a plurality of heterogeneous actors. It is in this sense that Rosenau speaks about the progressive rise of new spheres of authority (SOAs) that challenge the absoluteness of the nation-state as the sole locus of political power. Within SOAs, authority is relocated not only on the basis of actors' attributes but, more importantly, relationally (Rosenau 1999). Thus SOAs vary in form and structure depending on the type of authority they embody (command, bureaucratic, or epistemic) and on how hierarchically this authority is distributed (Rosenau 1999).[3]

Hence, when we analyze concrete episodes where global governance experiments are played out, we should observe critically who the actors are that have entered the game and what their action possibilities are. In this sense, in our analysis we should try to answers questions such as the following: Who are the actors that possess the knowledge resources necessary to manage global issues? What role do they occupy within the relational system among actors involved within global issues management? What kind and what amount of authority do they possess on the basis of their knowledge and of the exchanges they establish?

Interaction: What Kind(s) of Relations Are Actors Engaging In?

As mentioned above, the proliferation of political actors needs to be properly weighed according to the interactions established among them (Kooiman 2003). A first way to distinguish multiactor dynamics is to isolate the main level at which these dynamics happen. Indeed, the rearticulation of political arrangements is happening both at the national level (e.g., Peters 1995; Peters and Pierre 1998) and, what is more interesting to us, at the global level (e.g., Hewson and Sinclair 1999; Pattberg 2006; Rosenau 1995, 1999, 2002). Although the actual strategies enacted at the two levels will hardly be the same, whether the focus is directed toward transformations within the nation-state or between states, *fragmentation* should be considered the keyword and *recomposition* the antidote. It is in the tension between the proliferation of actors on the scene and the attempt to pull them together in a system created by their interaction that the dynamic element of contemporary governance processes lies. Indeed, "dynamics is decisive because of the nature and the direction of interactions involved" (Kooiman 2003:18). It remains an empirical question to determine which strategies for recomposition allow for an efficient managing of global policy matters. As suggested by Peters, the fragmentation of nation-states into different, more or less autonomous agencies is accompanied by several strategies aimed at reinvigorating state activity: the reproduction of market models; the enhanced participation of interested third parties; the removal of old constraints in order to foster social and political creativity; or the adoption of deregulation strategies (Peters 1995:293–313). Especially referring to supranational arrangements, but not exclusively in relation to these, the *network* idea has been often employed to suggest modes of reassembling dispersed units, thus providing an alternative to hierarchy and market models. However, as will be shown in more detail in the next chapter, different types of networks have been conceptualized and applied in the study of the national and supranational levels to better understand the interdependency that joins together institutional and non-institutional actors. Whatever the recomposition strategy or the coordination mode that will be chosen, it remains crucial to investigate the structural characteristics and the features of the final systems so as to identify actors who occupy more relevant or peripheral positions and to critically assess the consequences that different positions within networks have in terms of possibilities of action and power distribution.

Hence, while examining the relational element characterizing global governance processes, we should then ask questions like these: What kinds of relations do actors engage in? What kinds of resources do they exchange? What are the main characteristics of the resulting interactional system that is generated through resource exchanges (e.g., Is it a dense or a dispersed system? Is it fragmented or unified? Is it polarized or balanced?)? Who are the actors that occupy prominent positions within the system? What could be the reasons for their prominence?

1.3 The Multistakeholder Approach

One particular way of managing the complexity and thus taking advantage of the knowledge and resources possessed by the multiplicity of actors in the global context is to realize the multistakeholder processes (MSPs). These can be defined as processes that

> a) aim to bring together all major stakeholders in a new form of communication, decision-finding (and possibly decision-making) structure on a particular issue; b) are based on recognition of the importance of achieving equity and accountability in communication between stakeholders; c) involve equitable representation of three or more stakeholders groups and their views; d) are based on democratic principles of transparency and participation; and e) aim to

develop partnerships and strengthen networks between and among stakeholders. (Hemmati 2002:19)

The history of MSPs within global governance dates back to arrangements adopted by the International Labour Organization in 1919 in order to achieve a tripartite representation of governments, employers, and unions. Nevertheless, multistakeholderism became more known much later on, in 1992, after the United Nations Earth Summit (United Nations Conference on Environment and Development, UNCED) and the adoption of the plan of action known as Agenda 21 (Dodds 2002). This document identified nine major stakeholder groups (youth, women, indigenous peoples, non-governmental organizations [NGOs], local authorities, trade unions, business and industry, science and technology, and farmers) whose actual involvement was perceived as fundamental to reaching the goal of sustainable development. In general, United Nations (UN) activity has been often accompanied by different multistakeholder commissions and group experiments, of which the Commission on Sustainable Development, created in 1993, is probably the best known.

More recently, information and communication issues have become another global field, beside environment and health, where multistakeholder arrangements have been experimented with and strongly supported. In particular, the UN General Assembly, in conveying the World Summit on the Information Society (WSIS, Geneva 2003/Tunis 2005) through Resolution 56/183, encouraged "intergovernmental organizations, including national and regional institutions, non-governmental organizations, civil society and the private sector to contribute to, and actively participate in, the intergovernmental preparatory process of the Summit and the Summit itself" (United Nations General Assembly Res. 56/183:para. 5). Since then, the multistakeholder approach to global issues has known a revived interest, and reflections on the actual realization of multiactor processes have multiplied.

Overall, MSPs aim at bringing together for various purposes all key stakeholders that have a specific interest in a particular issue. In this sense, there are two main dimensions that determine MSPs' structure and organization: (a) the range of stakeholders included; and (b) the scope of the interaction (Susskind et al. 2003).[4] As far as the first dimension is concerned, it must be said that MSPs distinguish themselves from experiments such as deliberative democracy experiments (such as the collective and participatory definition of the community budget in Puerto Alegre; see Baiocchi 2003; Melo and Baiocchi 2006) because they do not involve all actors (potentially) affected by a certain policy but only representatives of the key stakeholders in the play (Susskind et al. 2003). The identification of key stakeholders in the play is then a crucial element that often cannot be managed in the best way because of structural limitations that hamper the recognition of "less obvious" stakes (Susskind et al. 2003:242). Thus, participants in an MSP can participate in their personal capacity because of their particular knowledge and expertise in relation to the matter under discussion, or they can act as representatives of an institution or of a non-institutional group and can commit to a certain constituency. The actual participants can be selected directly from the community they represent or by conveyors themselves on the basis of the scope of the process.

As far as the goals that can be pursued through a multiactor process are concerned, Susskind and his coauthors identify five categories of aims: relationship building; gathering and exchanging of information; agenda setting; brainstorming; and consensus building. Thus, the authors point out that considerations regarding the scope of an MSP are crucial as they provide the starting point from which to shape the very structure of the process as well as to set a threshold for the level of commitment of single participants. Accordingly, to achieve a shared and consensual provision it is necessary to involve representatives that can speak for a certain constituency, whereas, in order to build bridges between hostile groups or adversaries, the involvement of single committed individuals would be preferred.

There are several advantages in managing collectively common concerns through MSPs. Adam, James, and Wahjira (2007:8–10) identify a long list of them: a multistakeholder approach can promote inclusivity and equity in policy steering and implementing; it can expand analytical capabilities to address policy issues; it sustains grassroots mobilization and participation as well as the development of focused and holistic action plans; it can foster the sharing of skills and innovations; it facilitates training activities of new experts; it can create a balance between market-oriented and development-centered solutions; it can encourage good governance practices; it can enable participants to exploit their financial resources to good purpose; it can foster the creation of commitment both for leaders and for those individuals who are usually less active; it can promote and consolidate feelings of issue ownership and push for concrete action taking; and it can help to develop trust among groups that are usually hostile to one another.

However, the concept of multistakeholderism has been also criticized, not so much in theory but, more often, after any attempt to translate it into practice. Systematic reflections, especially during and after the WSIS process, have uncovered difficulties and incoherencies hidden under the rhetoric of participatory arrangements, inclusivity, and democracy enhancement. In particular, doubts have been raised in relation to the processes of identifying key stakeholders (Bendrath 2005; Cammaerts and Carpentier: 2004; Cammaerts and Padovani 2006; Padovani and Pavan 2011). According to the theory of MSPs, stakes can be defined following institutional distinctions that often result in the tripartition into governments, the private sector, and civil society (Susskind et al. 2003). This was precisely the case during the WSIS process. Alternatively, individuals can be free to decide to participate and under what constituency (personal capacity versus representative status). This is more the case of the Internet Governance Forum (IGF) process, in which interest in IG issues is a sufficient condition to be admitted to the discussion. However, it remains an open question whether these mechanisms actually ensure inclusiveness and the representativeness of all interests at stake. Moreover, in those MSPs that have been sponsored in the past by institutions such as the UN, it was quite common to consider different sectors of participants in monolithic ways, and this approach penalized in particular the heterogeneous universe made up by civil society entities as it minimized its inner variety and potential (Esterhuysen 2005). Other questions have been raised regarding the legitimacy and representativeness of the actors involved (Busaniche 2006; Cammaerts and Padovani 2006; Padovani and Pavan 2011). At present, especially in the management of information and communication issues, there are no systematic processes for delegating the responsibility of representing certain views and interests. Who is in charge of determining who will represent a certain stake? If self-selection, as in the IGF case, could raise the chances of achieving inclusive processes and thus overcoming formal representation logics, it also increases the risk of mistaking single opinions for collective statements and symbolic positions. Overall, then, "the normative values and democratic aspirations also tend to be easily lost in the translation of the concept [i.e., multistakeholderism] into concrete governance practices" (Padovani and Pavan 2011:546). In other words, the actual realization of MSPs hardly maintains the legitimacy requirement that is so clearly outlined at its theoretical level.[5]

In sum, rather than being an uncontested way of translating into practice global governance premises, multistakeholderism seems to be a "passe-partout concept" (Cammaerts and Padovani 2006:1), which has been differently understood and acted upon by different stakeholders in different occasions. In this sense, this specific concept is affected by the same vagueness that characterizes the discussion on global governance and the globalization process. But even though MSPs might not be perfect ways of managing the *diversity, dynamics,* and *complexity* of contemporary globalized contexts, they nevertheless generate experiments that deserve to be systematically analyzed, thus keeping in mind that what is behind the surface of a largely employed label is always the same: interdependency between a plurality of actors.

2. DISCUSSING INFORMATION AND COMMUNICATION ISSUES AT THE INTERNATIONAL LEVEL

Information and communication issues certainly stand at the core of contemporary global processes. Actually, as already mentioned, information and communication tools and processes are some of the fundamental forces driving the increased amount of interconnections and interdependencies among world areas. Consequently, their management is definitely of paramount importance and, given their degree of pervasiveness, might touch upon a plurality of interests. In this sense, this management requires the joint participation of heterogeneous actors, it stimulates the emergence of many claims for participation, and it can lead to the direct involvement of all stakeholders in decisional dynamics. Until now there have been two occasions for debating international information and communication issues. The first one dates back to the period between the middle of the 1970s until the middle of the 1980s and is known as the debate on the establishment of a New World Information and Communication Order (NWICO). The call for a new order in the field was raised within the United Nations Educational, Scientific and Cultural Organization (UNESCO) by newly independent countries gathered under the umbrella of the Non-Aligned Movement (NAM). The second and more recent occasion was the UN-sponsored WSIS, which was organized between 2003 and 2005 by the International Telecommunication Union (ITU) and which again brought information and communication matters into the spotlight together with the need for a reform of global governance arrangements in this field (and beyond it).

Observers noted that, although the WSIS process did not stem directly from the NWICO one, both processes actually constitute parts of an ongoing discussion in the GCG field, and an actual legacy exists between them (Padovani and Nordenstreng 2005; Pickard 2007; Raboy 2004). Also, both of these two debate occasions actually stand as meaningful historical precedents to the most recent official discussion experiment in the field: the IGF. This is not only a consequence of the fact that the IGF kicked off as an output of the WSIS process. More than this, there is a meaningful political continuity between the three events. The NWICO brought for the first time information and communication on the global scene, thus revealing that they constitute controversial sites where relevant power dynamics can take place. Using this framework, the WSIS provided new ground to experiment with innovative patterns of collaboration between different stakeholders, thus revealing how political collaboration can be highly influenced by actors' perceptions both of the matter under discussion and of the political mechanisms in which they are participating. In this connection, the IGF is broadening the margins for experimenting with multiactor interactions and the collective construction of meanings within an area of paramount importance in the GCG field: that of Internet management and development.

Yet observers noted also that the existence of an "events chain" joining together discussions from the 1970s with those deploying today is often over-looked. The denial of such a historical legacy was denounced in particular in the immediate aftermath of the first phase of the WSIS (henceforth, WSIS1). The main argument was that a sort of voluntary blindness was strategically adopted in order to hide deadlocks that could not be dissipated, probably because they stemmed from a set of broad and sclerotic inequalities innervat-ing contemporary world dynamics but that are preferably not made explicit. Conversely some observers noted that not only could more efficient political strategies derive from a deeper "historical depth in facing contemporary com-munication challenges" (Padovani and Nordenstreng 2005:265), but, more generally, grasping continuities and changes in the discussion (while keeping in consideration inevitable changes in the social, political, economic, and technological landscapes) could also help minimize the fuzziness of the *global governance* concept. Indeed, a longitudinal perspective allows for a compara-tive identification of opportunities and constraints to political action in a cer-tain point in time. In this sense, actors do not achieve the status of *winner* and *loser* once and for all, and the actual efficiency of governance arrangements can be evaluated contextually. In other words, wins and losses as well as the adequacy of adopted governance solutions are rather to be weighed in relation to power redistribution dynamics within given interactional contexts.

In line with this perspective and acknowledging that the GCG story is not told as often as it should be, the second part of this chapter provides a short review of these two debate antecedents to the specific process we analyze—the IGF. We would need a much more detailed analysis of these supranational processes if we tried to identify who won and who lost the game or if we wanted to systematically evaluate how efficient and effective governance arrangements adopted in the GCG field have been. However, as in this case, we just want to outline the main content and process elements that preceded the contemporary phase of discussion on IG; we will not dwell too much on the technical or organizational details of the two processes, and we will try to provide our readers with a complete and general overview.

2.1 The Call for a New World Information and Communication Order

The debate on the New World Information and Communication Order (NWICO) finds its roots in decolonization processes that occurred during the last century in many African, Asian, South American, and European countries. The political independence reached by ex-colonies resulted in the birth of new institutional actors that entered the international relations landscape as a "third bloc," as an alternative both to the Western bloc and to the socialist Eastern one, thus bringing into political arenas specific interests and needs. Indeed, having reached political independence did not mean that ex-colonies could enjoy a genuine independence from their former dominators. To promote an actual autonomy and avoid the creation of new, subtler dependencies, economic independence and cultural independence were the other two conditions that needed to be fulfilled. Because their achievement was not an easy task, a strategy of collective pursuit seemed to be the most effective one, and the so-called Non-Aligned Movement became the main platform sustaining the call for a New International Economic Order (NIEO)[6] and for a NWICO.

The link between economic and cultural matters and, consequently, between the call for a NIEO and for a NWICO was strong. In an overall context where information was becoming an increasingly crucial resource, "a lack of control over modern means of communication implied vulnerability not simply to cultural imperialism, but to yet another facet of outright economic domination" (Gallagher 1986:36). On the one hand, the debate on the NIEO served as a base to elaborate and strongly call for a new information order that was meant to be based on the same principles: self-determination, pluralism, participation, and cooperation. On the other hand, the NWICO not only was a fundamental element in the pursuit of more balanced economic relations but also could determine the actual establishment of a NIEO (Gallagher 1986).

The requests entailed in the NWICO proposal were complex and mainly based on general principles to be established in relation to all kinds of information, all media, and all kinds of communication technologies (Carlsson 2003). Concrete proposals to translate these principles into communication practices were actually missing, but, as Carlsson points out, this was mainly due to the difficulty of coordinating the suggestions to be applied in very heterogeneous local contexts. Overall, though, there were mainly four thematic "cornerstones" that recurred in the documents produced by NAM countries about a new information order, known as the *4Ds* (Carlsson 2003; see also White 1986 and Uranga 1986):

- The *democratization* of information flows, which had to be more balanced and equitable so as to foster accurate, accountable, and objective representations of newly independent people and territories;
- The *decolonization* of culture as an integral part of the independence process on the way to self-determination and the affirmation of ex-colonies' cultural identities;

- The *demonopolization* of transnational companies working in the information and communication field, thus supporting the promotion and the consolidation of national information systems;
- The *development* of society and societal systems under the light of self-determination and independence thanks to a more equitable distribution of information resources and the promotion of local journalism.

UNESCO provided the main stage for the NWICO debate by virtue of its commitment to "collaborate in the work of advancing the mutual knowledge and understanding of peoples, through all means of mass communication and to that end recommend such international agreements as may be necessary to promote the free flow of ideas by word and image" (UNESCO Constitution, para. IIa).

Advancements in the ICT field, especially after the Russian launch of the satellite in 1957, were definitely characterized in political terms, thus rendering the "free flow" issue another occasion to play out deeper controversies. Communication and information issues were indeed framed within an overall context of Cold War dynamics and struggles for independence, and, in this context, the NWICO debate with its North-South perspective ended up crossing the East-West line of conflict. Within the discussion, the Western bloc came to endorse a liberal position, concerned with the free circulation of information and defending the unbalanced status quo of world communications that functioned to preserve the work of big transnational companies in the field. Conversely, the Eastern bloc was a great supporter of state control over information and communication dynamics, particularly in opposition with the Western liberal model. Finally, the bloc of newly independent countries was struggling for greater roles in international forums, for a systematic reform of worldwide circulation of news and images, and for the development of local information systems that would promote a genuine independence from former dominators. When the Eastern bloc came closer to the non-aligned bloc (more because of envisaged possibilities to counteract its adversaries than for a real thematic and political proximity), Western countries feared a possible alliance against their interests. As a consequence of such a diffused mistrust, all attempts to reach an agreement on how to regulate the flow of information and communication systems were vain.

After this first period, which Uranga (1986) labels "political-ideological confrontation" and which lasted from 1973 and 1976, a new phase started with the Nairobi General Conference in 1976.[7] It was during this conference that the NWICO proposal was officially presented to the international community by the Tunisian information secretary of state, Moustafa Masmoudi. Thus the discussion did not lead to the definition of any agreement about the formulation of a document on the role of the media. The NAM position to post-pone the discussion until the following meeting prevailed over the general-ized dissent of the Western bloc on negotiations as well as on attempts of the Eastern bloc to immediately adopt a resolution. Also, in 1976 the so-called MacBride Commission—from the name of its president, Sean MacBride, Amnesty International's founder—was officially appointed to study commu-nication problems in view of the establishment of a NIEO and of a NWICO (Carlsson 2003).[8]

In the next 2 years, the discussion on the role of media continued and culminated in 1978 with the adoption of the *Declaration on Fundamental Principles Concerning the Contribution of the Mass Media to Strengthening Peace and International Understanding, to the Promotion of Human Rights and to Countering Racialism, Apartheid and Incitement to War,* better known as the "Mass Media Declaration"[9] (MMD). The MMD contained some gen-eral guidelines that provided an overall acknowledgment of instances pushed forward by ex-colonies about the necessity of a more equitable and bal-anced flow of information as well as about the need to foster economic and financial conditions to establish local and autonomous information systems and networks. Nonetheless, the adoption of the MMD exacerbated conflict dynamics between the East and the West. The belief spread among Western countries that the NWICO and the MMD were part of a Russian project to dominate information and communication: boycotting all the discussion was perceived as the appropriate strategy to isolate the Soviet Union and "the radicals" from the Third World (Lee 2003). However, despite the burden of these dynamics, the importance of the MMD for the articulation of informa-tion and communication issues should not be underestimated. Indeed, this document crystallized explicitly many of the claims coming from the devel-oping countries, and, for the first time, it framed these requests in the broader context of international law.

With the Belgrade General Conference in 1980, the NWICO entered a third phase of debate (Uranga 1986). In that year, the final report of the MacBride Commission was presented after 3 long years of work to provide an overall and systematic account of the problematic situation of information and com-munication flows. The report, *Many Voices, One World*, constituted the prin-cipal document of the NWICO debate. It comprised five parts[10] and made 82 final recommendations for actually translating a new information order into practice; however, it rarely pointed to concrete actions or to the actors that should be in charge of different tasks (Carlsson 2003). The General Confer-ence adopted the report with a resolution that encountered the firm opposition of the United States, the United Kingdom, Japan, Australia, and Canada. Even though the contents outlined in the document were reputedly accurate, they were not received as the basis for reforming worldwide communication systems. On the contrary, an alternative solution was set up through the adoption of a second resolution, namely, the one establishing the International Program for Development of Communication (IPDC). The IPDC was set up with four main goals: (a) *assistance* to the development of national communication systems; (b) *coordination* of international information exchanges and of interactions between UNESCO and the ITU, the regulatory UN agency for the information and communication field;(c) *information*-gathering activities on the needs of developing countries and on the status of cooperation among them; and (d) *fundraising* for carrying on IPDC tasks (Carlsson 2003; Padovani 1993).

As observers point out, the generalized attention conveyed on the program and the cold reception of the MacBride Report were a clear signal of the "new" perspective adopted by UNESCO on information and communication issues: one that was more oriented toward practical strategies and less to exhausting ideological conflicts (Carlsson 2003). These perspective and strategy shifts were perceived as politically necessary. While discussion on the NWICO was culminating, thus nurturing the hopes of the developing countries, the adoption of the MacBride Report was considered an aching flank for the Western bloc. European and North American media depicted the discussion on a new information and communication order as a threat to freedom of expression and to the free market while they accused UNESCO of being a politicized and biased organization working to promote radical projects (Lee 2003). Moreover, several doubts were raised about the legitimacy of the NWICO call itself, due to the massive repressions enacted by some of the NAM countries against internal dissent mobilizations. While internationally claiming the necessity to guarantee democratic information flows, some authoritative governments were locally denying this same principle, and this behavior inevitably jeopardized the credibility of the claims they were raising within UNESCO (Kidd and Rodriguez 2009). The fast decline of claims that had animated the discussion for more than a decade culminated in 1985 with the U.S. and U.K. withdrawal from the agency on the wave of preoccupation with the progressive loss of accountability of the organization itself (Lee 2003). After losing almost 30 percent of its funds and suffering from the general mistrust surrounding its actions, the organization adopted an apparently neutral perspective according to which technological and infrastructural improvements were considered as the best solution to balance world information fluxes (Padovani 1993). The IPDC became the emblem of the new strategy that was formalized in the 5-year plan *Communication in the Service of Men* and that became from that moment on the institutional cornerstone for resolving worldwide communication imbalances.

2.2 The World Summit on the Information Society

After the fall of the NWICO debate from international attention and from international political arenas, information and communication issues continued to be managed by UN agencies and organizations yet without being under any particular spotlight. By the latter half of the 1990s, though, when it appeared evident not only that information and communication matters had remained central but also that their relevance in contemporary globalization processes was higher than ever, the interest in them revamped and a necessity to resume the discussion emerged rapidly. In a sort of continuity with the discussions deployed in the 1970s and the 1980s, UNESCO was initially identified as the proper host organization for the next "Conference on Information and Communication for Development" whose preparation had been undertaken by its Executive Council in 1996 (Ó Siochrú 2004a). The conference agenda was meant to be very comprehensive: indeed, it had to address all thematic areas within which the role of information and communication processes was marking a significant contribution for global development. Furthermore, the conference was intended to be comprehensive in terms of its participation, as all actors that had a stake in this field had to find their place within the discussion. However, a few months after the beginning of its preparation, the process was interrupted and the conference project was aborted for no explicit or clear reasons (Ó Siochrú 2004a).

In 1998, it was the ITU that picked up again the idea of realizing a WSIS. This shift from UNESCO to ITU was not just a matter of formality but entailed substantial consequences. While the former organization had a long-term tradition of discussing information and communication matters in relation to society and culture and had a marked tendency to collaborate with non-governmental entities, the latter had a narrower agenda comprising basically technical and infrastructural matters and was lacking any experience in promoting collaborative efforts with non-governmental actors (Ó Siochrú 2004a). Nonetheless, in a context where globalization processes were progressively fostering the emergence of non-state actors onto the political scene and where levels of trust and confidence in traditional state-managed summits were rapidly diminishing (Hintz 2009; Ó Siochrú 2004a), it would not have been possible to realize a solely state-centered meeting on issues that were so globally and socially relevant. In this sense, "the WSIS process was situated at a strategic crossroads between an old model of an exclusive government sphere of decision making (with some elementary but strictly limited attendance of non-state actors) and a new model on participatory governance" (Hintz 2009:118).

Therefore, ITU elaborated a format for the WSIS that was innovative in many aspects (Klein 2004).[11] The first innovation consisted in the double-meeting formula adopted: while traditional summits are usually "one-time events," the WSIS would have gathered in 2003 in Geneva and in 2005 in Tunis (International Telecommunication Union Res. 1179/2001; United Nations General Assembly Res. 56/183). Also, the UN General Assembly Resolution 56/183 stated that each of the two meeting had to be preceded by an "open-ended intergovernmental preparatory committee" (para. 2). Each preparatory process (PrepCom) was also paralleled by regional conferences aimed at bringing at the global level the largest number of inputs and visions developed locally.[12] Actually, the majority of interactions and negotiations occurred during the PrepComs rather than during the official meeting days. The Geneva and the Tunis meetings ended up being mainly "ceremonies of ratification" (Klein 2004:5) where dynamics previously deployed did crystallize in the adoption of the final documents. Thus, the double-meeting formula translated into an unusual length of the process (almost 5 years) that was actually appropriate considering that the Summit was aimed at "the development of a common vision and understanding of the information society and the adoption of a declaration and plan of action for implementation by Governments, international institutions and all sectors of civil society" (Res. 56/183, Preamble). Such an ambitious goal could be hardly achieved, if at all, without ensuring a suitable time span.

The second innovation was perhaps even more important. The UN General Assembly encouraged "other intergovernmental organizations, including international and regional institutions, non-governmental organizations, civil society and the private sector to contribute to, and actively participate in, the intergovernmental preparatory process of the Summit and the Summit itself" (Res. 56/183, para. 5). Not only was the WSIS pursuing an ambitious task, but it was also meant to be innovative in the procedures used to fulfill its commitment. The Summit then became the latest experiment for multistakeholder collaboration in global governance and an unprecedented attempt to enlarge participation in the field of information and communication issues, although the roles and responsibilities of governments, the private sector, and civil society were not outlined in any documents or guidelines. This major call for collaboration fell in a context characterized by a widespread procedural uncertainty: while the UN encouraged contributions and active participation from all stakeholders, it foresaw a direct responsibility of *inter-governmental* PrepComs in defining the agenda of the Summit, in finalizing the draft Declaration and the draft Plan of Action (the two final documents of the first phase), and, more importantly, in deciding on the modalities of the participation of other stakeholders (Res. 56/183, para. 2). It is not clear if the reasons underpinning the choice of promoting a multistakeholder summit are to be taken back to ITU ignorance of conventional summit mechanisms or to a genuine will to reform governance arrangements (Ó Siochrú 2004b). For sure, the fact that the final Declaration and the Plan of Action had no binding character lowered the risk of explicitly supporting joined participation. What is clear, instead, is that attempts to reduce procedural ambiguities constituted the very core element of the Summit, actually reducing the possibility of confrontation on content that was highly dependent on participation mechanisms outlined for the three sectors.

Different stakeholder groups participating to the WSIS (and single constituencies within them) had obviously different agendas and priorities for the information society. Some of them did not even agree on the term *information society,* since they believed that it carried with itself a tradition where the liberalization of markets, the privatization of services, and the mere infrastructural element were obscuring the social and human-oriented side of the technological revolution, thus conveying a particularly static vision of society itself (Ó Siochrú and Girard 2003). On these bases, inviting governments, intergovernmental institutions, the private sector, and the vast universe of civil society organizations to officially take part in the same institutional process entailed much more than what conveners could expect: it was opening the doors for a confrontation between identities, visions, and values that often were in contrast to one another, and it meant experimenting with conflictual dynamics that were deploying at the symbolic level but that were not less severe than real contentious political manifestations, as shown in the next chapter.

Issues that were brought to the table were very numerous, even though initially the ITU did not actually mean to face them all during the Summit (Ó Siochrú 2004a). The content richness that characterized the discussion on the WSIS1 phase is shown by the multiplicity of thematic groups (caucuses or working groups) within which the civil society sector organized itself to better channel competencies and ideas: media and community media; cultural and linguistic diversity; values and ethics; e-government and e-democracy; health and ICTs; patents, copyright, and trademarks; privacy and security; Internet governance; volunteering and ICTs; persons with disabilities; scientific information; education and academia; the relationship between environment and ICTs; human rights; gender issues; indigenous peoples and ICTs; financing mechanisms; young people and ICTs; and cities, local authorities, and ICTs. Over this vast range of issues, many different points of views crossed one another: institutional preoccupations for protecting national interests; private sector worries ultimately linked to the generation of profits and to industry flourishing within globalized markets; and civil society's myriad perspectives, sometimes oriented toward the defense of the interests of specific categories, sometimes articulated in relation to broader dynamics and situations. During the WSIS1 process, some topics—financing mechanisms, Internet governance, intellectual property rights, human rights, media and security issues—emerged as particularly problematic (Hintz 2009). The Tunis meeting, then, was thematically centered on these unresolved nodes and, in particular, on two of them: financing mechanisms and IG. As these proved to be particularly controversial, two specific groups were set up at the end of the Geneva Summit to prepare the terrain for negotiations in Tunis: the Task Force on Financial Mechanisms (TFFM) and the Working Group on Internet Governance (WGIG). Discussions on more human rights and social issues were instead hindered by the hostile behavior of Tunisians, especially within the civil society sector, up to the point that the second and the third PrepCom to the African meeting had to be moved back to Geneva in an attempt to minimize political interferences during negotiations (Hintz 2009).

Clearly, the actual opportunities provided to the three sectors to convey their inputs to the first preparatory process had a strong influence on the very shape of the "information society vision" resulting from the WSIS1 process. Much of the WSIS1 discussion actually focused on procedural rules for translating the principle of a tripartite approach to discussion into practice.[13] In fact, it was definitely hard to find a balance between the hopes of realizing a real multistakeholder dialogue and the difficulties of realizing a truly participatory process. In this sense, the WSIS became not only the locus where a discussion about some of the most relevant global issues was developed but also the very occasion to test a twofold set of collaboration mechanisms: on the one hand, mixed dynamics involving governmental and non-governmental actors; on the other, internal dynamics deploying between actors gathered in the same sector (e.g., between North and South of the world governments, local and transnational industries, and the heterogeneous components of the civil society sector). This latter mechanism was particularly relevant for the civil society constituency, which was holding together an enormous number of different organizations, varieties of expertise, and visions that had never come together before and that, until then, were never offered the possibility of systematically interacting with institutional actors within institutional occasions (Ó Siochrú 2004a).

Difficulties in finding an effective place within the WSIS1 preparatory process forced civil society to internally organize so as to facilitate cohabitation with the private sector and the governmental group within the overall WSIS institutional framework.[14] Even after a very harsh beginning in PrepCom1 before the Geneva meeting, civil society continued to try to find its way to actively contribute to the shaping of a common vision on the information society. However, it also continued to be allotted short time slots for illustrating broad and complex concerns, while it remained excluded from the negotiations and actual decisional dynamics that were dominated by governmental actors who remained solely in charge of proposing and voting on the contents of the two final documents (Ó Siochrú 2004b). By the time of the third PrepCom before the Geneva meeting, civil society had certainly tried to exploit at the top the few opportunities at its disposal to contribute to the Draft Declaration. It had established a formal structure (i.e., the Civil Society Bureau) that facilitated its interaction, especially with governmental actors, and that was accepted by all of its members as "legitimate enough" to negotiate with other stakeholders groups (Ó Siochrú 2004b). Still, the effort did cost a lot, and the rewards received were not considered enough. In the light of the little consideration that its inputs had received and to avoid a definitive penalization of its view in the final Declaration, at the end of PrepCom3A in 2003 civil society officially withdrew from the drafting process, suspending its inputs and focusing on the elaboration of its own document, *Shaping Information Societies for Human Needs*, which was added as a further final contribution during the Geneva meeting (Hintz 2009; Ó Siochrú 2004a, 2004b). At the end of the WSIS1, opposite evaluations clashed on the results achieved: some observers underlined that a separate declaration and the limited impact of civil society on the final documents represented the ultimate disappointment of the multistakeholder promise, whereas others emphasized that civil society's internal organizing process was one of the most valuable outcomes of the whole process (Ó Siochrú 2004b; Raboy 2004).

The Tunis phase of the process (WSIS2) was of little help in deciding which of these two positions was the most adherent to truth. In Tunis there was a more limited range of issues to be discussed (mainly Internet governance and financing mechanisms) because of a natural focus on the most urgent matters that emerged in WSIS1. Also, the WSIS1 controversy over procedures had consumed so many energies from the participants that, overall, interactions between stakeholder groups were less harsh (although not peaceful). During PrepCom1 in Hammamet it was agreed that the focus of the WSIS2 would have been threefold: (a) on the implementation and the follow-up of the whole WSIS process; (b) on the results of the work by the (TFFM); and (c) on the results of the work by the WGIG (CONGO and NLGS 2005; Hintz 2009). As mentioned above, the TFFM and the WGIG were constituted at the end of the Geneva meeting as multistakeholder groups aimed at fostering the discussion on the two most controversial issues that had emerged: how to finance ICTs for development and Internet governance mechanisms. A more detailed elaboration on WGIG's work will be provided in the third chapter. For now, it is enough to say that the way in which the two "special bodies" were created and the procedures followed to represent the different stakeholder groups within them had been heavily criticized. This was the case especially of the TFFM, while the WGIG, in general, was taken in some cases as a good example of multistakeholder collaboration that was kicked off and managed in a transparent way, thus ensuring peer representation of various positions (Currie 2005; Hintz 2009). Still, during the Tunis preparatory meetings, civil society found itself constrained by strict procedures for contributing to the two final documents (the Tunis Commitment and the Tunis Agenda) and, in the end, developed its own document, *Much More Could Have Been Achieved,* through an online steering process. In this document, civil society recalled its commitment and its efforts to contribute positively to the whole WSIS process but acknowledged that more could have been achieved under more favorable participation conditions (WSIS Civil Society, 2005). Despite several disappointments derived from the Geneva and the Tunis experiences, civil society groups recognized the importance of participatory arrangements and declared that their future efforts would have gone toward the next phase of discussion, namely, the IGF.

3. COMPARING OFFICIAL INTERNATIONAL DISCUSSIONS: CONTINUITIES, CHANGES, AND CHALLENGES

In concluding this first chapter, we try now to join together the two strands of reflection on the structural and historical elements of GCG with the aims of identifying the continuities and the changes in the field and of designing the contours of the situation in which contemporary discussions on IG started. To better exploit the inherent potential of a longitudinal perspective to reconstruct the GCG field, continuities and changes will be referred to the two categories of contents and processes. What we obtain from the intersection of these two analytical dimensions, then, is a fourfold chart that can help us grasp the main elements that should be addressed through actual applications of the analytic framework proposed in this work (see Table 1.1).

Looking at contents, a first element of continuity that we find in discourses developed during previous occasions of debate is the argument that "media are profoundly essential to the fulfillment of human needs and the realization of human dignity in the modern world" (Calabrese 2004:325). While this "developmental idea" constituted basically the main topic in the NWICO proposal and was sustained in particular by the NAM, during the WSIS it was carried on mainly by the civil society sector. Also, in the two processes this argument served different purposes: in the NWICO case, it was used to achieve a genuine independence from former dominators also through a more balanced communication world order; during the WSIS process, it

Table 1.1. Continuity and change elements in relation to GCG content and process

	Continuity	*Change*
Content	- Developmental perspective on the application of ICTs to overcome divides - Need for regulation - Neoliberal argument	- Agenda enlargement - Agenda refinement
Process	- Centrality of institutional actors - Lack of mentality change	- Inclusion of non-institutional actors in the official process - Mobilization around information and communication issues

GCG = Global Communication Governance; ICTs = Information and Communication Technologies.

was used to contrast the predominance of technical and commercial visions guiding the development of contemporary societies and to promote more inclusive and democratic supranational arrangements. Empirical analysis of documents drafted during the two different periods confirms the duration of this developmental idea across processes, thus paralleling the centrality of Information and Communication Technologies in society with the challenging task of designing appropriate regulatory mechanisms for them (Padovani 2005a:327–330).

There is a second continuity thematic element that stands in a dynamic relationship with the developmental thesis, namely, the neoliberal argument. Pickard (2007) argues that both the NWICO and the WSIS processes provided two spaces for articulating the dichotomy between the neoliberal argument and the responses to it. Accordingly, the call for a NWICO should be seen also as an answer to the "neoliberal imperatives advanced by a powerful Western state-corporate alliance" (Pickard 2007:119), while its defeat should be read in strict connection with the prevailing of these same imperatives over the requests for the democratization of media systems (as also suggested by Carlsson 2003 and by Mowlana 1993). In the same way, most of the civil society efforts during the WSIS process could be thought of as a response to the predominant market logic that was put in place by regulation processes on ICTs and privileged international-aid mechanisms based on mere commercial and technological transfers (Carlsson 2007). According to this perspective, the symbolic contraposition pushed forward during the WSIS between the pluralistic idea of "communication societies" and the monolithic "information society" (Ó Siochrú 2004b; Ó Siochrú and Girard 2003) is emblematic of the dynamic relationship existing between the developmental and the neoliberal arguments.

Thus, agendas of discussion have changed over time. The technological evolution and its capillary pervasiveness are mirrored in the proliferation of issues that have entered the discussion in the period between the NWICO and WSIS processes. The different topics debated during the NWICO—such as the need for objective news reporting and for the establishment of independent, South-centered press agencies—were framed within the overall idea of the *4Ds* (democratization, decolonization, demonopolization, development) that constituted the main thematic cornerstones in the discussion. At the end of the first WSIS phase, the agenda of discussion was much more "plural," as it included infrastructure, access, capacity building, trust and security, enabling environments, ICT applications, cultural and linguistic diversity, and the ethical dimension of the information society (Padovani 2004:189). The actual coexistence of a multiplicity of thematic strands that were nurturing the discussion is verified by textual analysis of documents produced during the WSIS1 (Padovani and Tuzzi 2004).

However, it is interesting to notice that during the WSIS 5-year process this agenda enlargement was followed by its progressive refinement. The emergence of IG and, to a lesser extent, of financial mechanisms for ICT development as the two main discussion topics after Geneva constitutes a "change in the change." Thematic focalization certainly responded to a substantial urgency, as Internet and financing mechanisms were actually the two priorities within many stakeholder plans. Nonetheless, agenda refinement should also be interpreted as a consequence of the link between contents and processes in the GCG field because, as Selian points out, "the fact that the scope of interests, the range of participation, and even the basic terminology at the core of the WSIS are so broad and nebulous does not lend well to the much-needed process of prioritization" (2004:213). In other words, the thematic proliferation was jeopardizing the development of the process itself and was consequently reduced to minimize the risk of ineffectiveness of the whole process. Changes in contents should be seen then as the byproduct of two intertwined dynamics: the evolution of the field itself (i.e., technological innovation and innovation in the uses of technology) and the evolution of practices in the field (i.e., innovation in processes of discussion and policy making).

Looking more closely at the dimension of process, a relevant trait of continuity is provided by the central role that is occupied by institutional actors in the overall development of the field (Cammaerts and Padovani 2006; Selian 2004) and that is justified by the early awareness that an efficient management of information and communication issues requires supranational arrangements (Padovani and Pavan 2011). But if this centrality is hardly surprising in the NWICO context—where the rules of the game were those of traditional international diplomacy and, therefore, revolved around state actors—it becomes more interesting in the WSIS—where it was challenged by the presence within negotiations of private sector and public interest initiatives. In the WSIS case, despite expectations tied to the formal multistakeholder feature of the process, the centrality of governments and of institutional actors in general was ultimately guaranteed by the summit format that was adopted, which privileged some institutions over other participants, thus guaranteeing them and their allies wider margins for maneuvering (Klein 2004). The formal enlargement of participation opportunities provided in the WSIS was constantly accompanied by an informal (but effective) strategy that consisted of considering non-institutional inputs at the lowest level possible and that was backed by the definition of strict rules of procedure. Overall, observers agree that while formal windows of opportunities were opened up, there was little room for newcomers to substantially influence the official negotiation process.[15]

The unmodified centrality of states in the dynamics we examined points toward a second element of continuity in the GCG processes: the lack of a "mentality change" that would be instead required to actually innovate global diplomacy (Padovani 2005a). Such a lack crystallizes in the definition of roles for institutional and non-institutional actors that are hardly subjected to change over time and do not respond to the structural modifications that occur at the global level. On the side of governments, this sort of blindness translates into explicit attempts to preserve strong powers and maximize internal advantages; on the side of non-institutional actors it can take different forms: from a careless attitude characterizing the business sector (Ó Siochrú 2004b; Padovani 2005b) to the difficulty of civil society entities to perceive themselves as "part of the process." If international diplomacy settings in the NWICO period did not allow for any involvement of civil society in the official discussion and strengthened its self-organizing attitude (outside institutional forums), evident difficulties in the realization of actual multiactor collaborative dynamics during the WSIS process reinforced the civil society tendency to set up non-institutional and alternative initiatives at the margins of the official process (Hintz 2007, 2009; Milan 2006), thus channeling inputs into autonomous declarations or statements (Klein 2004; Selian 2004).

Despite the disappointments, frustrations, and difficulties experienced over the past 20 years, sparks of change are illuminating a possible path toward the reform of supranational politics starting from the GCG field. The most visible element, in this sense, is provided by the formal involvement of non-institutional actors, in particular, of civil society entities within political processes defining the GCG field. As mentioned, the NWICO did not formally involve civil society, with the sole exception of sporadic participation of some NGOs with consultative status in surveys promoted by UNESCO. The burden of articulating a counter altar to the Western neoliberal position was all on the

NAM's shoulders. This formal exclusion from institutional debate does not mean that critical positions were not being articulated outside UNESCO or that claims were totally ignored during the official debate. Several non-institutional realities multiplied over the years of the NWICO debate or in its immediate aftermath in the attempt to widen and translate into practice the principles that had guided the request for a new world information order.[16] Even though the protagonists of the NWICO debate knew these initiatives, they developed and remained outside the loci of official discussion and had no voice within UNESCO meetings. The formal involvement of civil society in the Summit marked instead a neat change in this sense, as non-institutional inputs were recognized (at least in principle) as a legitimate part of the effort to construct a common vision and an action plan for the information society. Shortcomings of the WSIS experience and the general dissatisfaction of civil society participants with the actual dynamic of negotiation can be basically seen as the result of a disparity between initial expectations for agency possibilities and the difficulties of actually sharing power among the different actors involved, worsened by an overall undertheorization of multistakeholderism itself (Cammaerts and Padovani 2006). Rather than offering a model for realizing multiactor processes, the WSIS crystallized the need to design appropriate solutions for an inevitable cohabitation between institutional and non-institutional actors (Hintz 2009). Therefore, it should be considered not as an arrival point but quite the opposite: as a learning experience (Cammaerts and Padovani 2006).

Nonetheless, a relevant modification was also experienced by the civil society sector itself as a consequence of the participation model entailed in the WSIS structure. In this sense, the WSIS offered civil society entities five main opportunities: (a) it provided the opportunity to make different civil society initiatives converge and create synergies; (b) it offered a world stage to make their claims and actions visible; (c) in forcing its internal organization process, it increased cohesion between the various parts; (d) it fostered consensus-building activities through collective drafting of documents and the clarification of concepts; and (e) it promoted reflections on issues of participation, accountability, and legitimacy (Padovani and Pavan 2009a:224). Informal outcomes of broader participation arrangements led to a systematic reflection on the existence of a structured mobilization on information and communication issues (Hintz 2007, 2009; Milan 2006; Padovani 2005b; Padovani and Pavan 2009b), sometimes understood in terms of a real and structured "global movement" (Calabrese 2004). For sure, the rediscovered legacy between the NWICO and the WSIS processes inspired some observers to borrow the idea of *movement* to translate at the level of the process those thematic continuities referable to the dichotomy between the neoliberal and the developmental arguments we pointed out before. However, when the *movement* idea was applied to the mobilization of civil society groups within and beyond the WSIS occasion, it lost the institutional connotation it had when it was referred to the non-aligned countries. At the beginning of the new century, the movement on information and communication issues was intended as a comprehensive gathering of all those civil society initiatives that, from the middle of the 1980s on, had continued to pursue the efforts of democratizing communication systems, thus translating the NWICO principles directly into society and daily communication practices. In this perspective, the WSIS had provided the opportunity for finding a middle ground between a multiplicity of experiences, agendas, visions, and ideas. While in the past initiatives such as the MacBride Roundtables, the People Communication Charter, the WACC (World Association for Christian Communication), the AMARC (World Association of Radio Broadcasters), or more extreme experiences such as Indymedia worked for the same scope but disjointedly, the WSIS led them to convergence and mediation. The fact that a wider mobilization on information and communication matters existed "out there" (Padovani and Pavan 2009a:224) facilitated the reach of a shared position during the WSIS and fostered the consolidation of a discourse on communication and information issues that was not only technical but also more socially oriented. Overall, then, the disenchantment regarding the multistakeholder approach to global politics fostered a progressive shift of attention focus: from the search for changes in the formal process toward the search for changes in the dynamics of mobilization around communication and information issues on the side of civil society groups.

However, already on the way to Tunis but, more significantly, after the end of the WSIS process, the idea that a *movement* on communication and information issues was emerging cooled down. Furthermore, while acknowledging that supranational political dynamics on information and communication matters continue to be characterized by *diversity* (of actors involved and themes brought into the GCG domain), *dynamics* (leading to incremental changes in the conduct of supranational politics), and *complexity* (of interactions taking place between actors and themes) (Padovani and Pavan 2008), the usefulness of the multistakeholder concept for studying how the GCG field was consolidating started to be questioned. Consistently, "more articulated approaches to understand how the different normative assumptions of actors translate into practice" (Padovani and Pavan 2011:546) have been pushed forward to analyze multiactor interactions but also to look more specifically into mobilization dynamics on information and communication issues. In the end, the WSIS experience revealed to practitioners and researchers that something was actually happening, that politics were changing and that governance arrangements were moving toward the inclusion of previously excluded and external actors.

Still, the idea of multistakeholderism, both in the political practice and as a theoretical lens for research, scratched only the surface of multiactor processes: diversity was considered only with specific reference to actors' attributes, and the scope of multiactor dynamics was evaluated mainly in terms of impacts on official political outputs. What are then the critical elements emerging from our overview of the GCG field and that challenge the explanatory potential of the concepts of multistakeholderism and that would need to be dealt with through innovative approaches and perspectives of research? Let's try to answer this question by using again the categories of contents and processes.

As far as contents are concerned, most of the reflection so far has been focused on the *presence* or *absence* of determined thematic inputs within supranational processes' official documents. In this sense, the impacts of civil society participation in the WSIS1 phase were measured starting from the presence or absence of some concerns from the final declaration and the plan of action. Much less has been written about the effects of the joint participation of different sectors on the progressive shape of a multiactor discourse beyond formal process outputs. Although the presence/absence dimension grasps well the *plurality* of global issues, it seems less capable of also accounting for the *dynamics* and *complexity* that characterize global politics. In looking at how multiactor dynamics are developing in the GCG field, we should definitely try to address some general questions that point directly to the collective construction of controversial matters, such as IG. In this sense, in approaching the study of how multiactor dynamics are shaping a new IG discourse, we should try to answer questions like these: What are the elements that belong to the IG agenda today? Has the agenda broadened or restricted passing from the WSIS to the IGF? Is it possible that IG is providing only the overall "thematic box" where multiple discourses are converging into and where new discourses are created? And if the thematic richness that has emerged together with the WSIS process is not canceled but has moved toward the IGF, has it been influenced by the technicality characterizing the Internet governance area? If so, in which way?

To answer all these questions, it is paramount to map out the agenda in the IG area by looking not only at what themes are included and excluded but also at connections between items of the political agenda. The thematic richness that emerged during the WSIS as a consequence of the direct involvement of civil society seems to have converged along the institutional supranational path leading to the IGF, but it might have undergone some modifications that need more clarification. However, in order to evaluate how the political agenda has modified and how continuities and changes impact the formation of a new IG discourse, listing present or absent themes is not enough. In fact, it is necessary to examine how themes are linked to one another, and how they are put in relation to form new challenges that stimulate the interest and require the competencies of a plurality of actors, both institutional and non-institutional . Therefore, shifting from a perspective of analysis that is solely centered on the presence or absence of thematic inputs toward an approach that looks also at how themes, concepts, and thematic strands are related to one another is a necessary step if we want to truly understand how multiactor dynamics contribute to frame and shape collectively meanings and understanding of global issues.

This latter point ties directly to challenges pertaining to the process dimension. It has been shown that, at the level of the formal process, traditional mechanisms that have been regulating international relations and supranational arrangement tend to persist. In other words, states continue to be the main protagonists in the game. However, in this case the sole evaluation of the presence or absence of actors within institutional processes does not allow a proper understanding of the consequences and relevance of the multiactor dynamic. So far, it has been extensively pointed out that the role of civil society as a "young stakeholder" consists mainly in its watchdog role and in its provision of knowledge to institutional actors (Padovani 2005b; see also Ó Siochrú 2004b). But this role was matured in a context where opportunities for participation were only formally extended while substantially supporting traditional political protagonists. How are the roles and functions of different actors' categories redefined (if at all) within institutional contexts that are explicitly set up to confront opinions and visions? Probably mindful of difficulties encountered during the WSIS process, states and intergovernmental institutions agreed to realize new experiments like the IGF and to further reduce barriers to access for all interested parties. Still, we cannot deny that the enlargement of participation to the definition of information and communication issues comes from a real awareness of civil society's fundamental and necessary expertise.

How and when does expertise translate into political leverage, if at all? If political opportunity structures, understood as "consistent—but not necessary formal or permanent—dimensions of the political environment that provide incentives for people to undertake collective action by affecting their expectations for success or failures" (Tarrow 1994:85), are being developed to exploit to a larger extent the potential of multiactor joint activity, are political mechanisms flexible enough to hold safely to the resulting diversity, dynamics, and complexity of political interaction? If the sole provision of chances to undertake action (a joined multiactor action, in this case) will not automatically translate into concrete participation (Koopmans 2004a, 2004b; Kriesi 2004; Meyer 2004), what are the incentives that foster the participation of institutional and non-institutional actors in the same dynamics? Are the bases of political collaboration and the roles of actor being redefined thanks to the opening up of innovative windows of opportunity? If so, in which way? When formal obstacles to participation in official dynamics are removed and the relations established become more important than simply "being there," what are the logics sustaining interactional patters between different involved actors? Is the provision of innovative and open political opportunity structures enough to finally trigger a chain reaction that will actually reform supranational politics?

If a perspective centered on multistakeholderism left relations between themes and actors behind, we propose instead to adopt ties as the starting point for our analysis. As the WSIS experience showed, indeed, practical realizations of multiactor dynamics detached quite meaningfully from their premises, and yet they have inevitably produced a change in the conduct of supranational politics. If we want to understand the directions that supranational politics are taking and how information and communication will be regulated in the near future, we need to answer all the questions we listed above, but we stand in a context where traditional conceptual tools that were developed to depict the complexity of political interactions have proved to be unsatisfactory to this aim. For these reasons, starting with the structural and historical elements of the GCG field, we designed a specific analytic framework that allowed us to account for contents, process, and their interplay according to a relational perspective.

NOTES

1. Held et al. identify four types of impacts: (a) decisional: cost-benefit evaluations that are influenced by global forces and conditions; (b) institutional: the extent to which institutions and organizations reflect the choices that are made available by global processes; (c) distributive: how power is redistributed among different actors; and (d) structural: the extent to which domestic social, economic, and political organization and behaviors are structured according to global conditions (1999:16–21).

2. In his review focused on challenges faced by the British government, Rhodes (1996) identifies six different uses of governance: as the minimal state; as corporate governance; as the new public management; as "good governance"; as a sociocybernetic system; as self-organizing networks. More recently, and building on previous elaborations, Pattberg (2006) identifies 10 meanings of the term *governance*:

as the minimal state; as the emergence of new commercial management strategies in the public sector; as corporate governance; as "good governance"; as coordination and cooperation efforts in decentralized networks; as multilevel governance within the European Union; as the second term of the legacy governance-govermentality elaborated by Foucault; as participatory governance; as socio-cybernetic steering processes; as that international order in international relations that focuses on international institutions and regimes coping with transnational problems.

3. The sphere of authority idea was originally developed to analyze international relations but could also be applied to internal dynamics whereby governments try to complete their tasks with the help of other actors who possess key resources (e.g., expertise, new models of organizations, etc.) that are necessary for respecting an agenda that exceeds their capabilities.

4. In their work, Susskind et al. refer to multistakeholder *dialogues*, that is, multiactor meetings that imply face-to-face interaction based on dialogue (2003). Thus, they acknowledge that multistakeholder processes provide a broader category that includes also multiactor collaborations that are not based on actual dialogue. Therefore, their elaboration refers to immediate interactions between key stakeholders, but we can nevertheless expand their considerations also to the broader category of multistakeholder processes as we conceive dialogue as a metaphor for social interaction among political actors (see Introduction). For this reason we will refer to multistakeholder processes (MSPs), adapting original points made by authors while speaking about *dialogues*.

5. "MSD [i.e., multi-stakeholder dialogues] designers must work to generate and maintain legitimacy from beginning to end" (Susskind et al. 2003:254).

6. The proposal for a New International Economic Order (NIEO) was agreed upon by ex-colonies during the 1973 NAM conference that was held in Alger and was then officially adopted in 1974 by the United Nations General Assembly (Res. S-6/3201). The UN resolution asserted that the establishment of a NIEO should be a general concern to be urgently addressed so as to reach a global balance within economic exchanges "based on equity, sovereign equality, interdependence, common interest and cooperation among all States, irrespective of their economic and social systems which shall correct inequalities and redress existing injustices, make it possible to eliminate the widening gap between the developed and the developing countries and ensure steadily accelerating economic and social development and peace and justice for present and future generations" (Preamble). In sum, the NIEO proposal was intended to promote a renewed structural balance that could advance the position of the newly independent countries in relation to industrialized states on the bases of five main requests: (a) better terms of trade for the Third World; (b) more Third World control over productive assets; (c) more South-South trade relations; (d) more Third World counterpenetration; and (e) more Third World influence on world economic institutions (Galtung 1986:10–11).

7. The first phase identified by Uranga can actually be dated back even to 1972, when, during the General Conference, the adoption of the *Declaration of Guiding Principles on the Use of Satellite Broadcasting for the Free Flow of Information,*

the Spread of Education and Greater Cultural Exchange was the engine starting the opposition between the Western and Eastern blocs. The declaration explicitly introduced the necessity of state agreements for direct broadcast communications and was actually proposed by Russia; it was indeed perceived by the Western bloc as a threat to the freedom of information and, above all, to the consolidation of transnational broadcasting companies (Hamelink 1994).

8. The MacBride Commission was meant to be composed to ensure pluralisms, representativeness of different cultures and socioeconomic conditions, and competence in the field. Yet it reflected the overall conflictual environment within which the project was taking shape: indeed, China refused to participate in the commission's work (Carlsson 2003; Lee 2003).

9. In fact, the Mass Media Declaration was adopted with 61 votes coming from the NAM and the Eastern bloc, one contrary vote (Switzerland), and the total abstention of Western countries (Hamelink 1994).

10. The five parts were (1) Communication and Society; (2) Communication Today; (3) Problems and Issues of Common Concern; (4) The Institutional and Professional Framework; (5) Communication Tomorrow.

11. ITU Res. 1179/2001 states: "The Summit meeting is to be organized under the auspices of the United Nations Secretary-General with the ITU taking a leading role in its preparations in cooperation with interested U.N. and other international agencies and the host countries." Although according to the documents the ITU played only "a leading role" in the process, we will refer the preparation process and the whole management of the Summit to it, as, in substance, the ITU was guiding the organization of the WSIS itself.

12. The WSIS1 in Geneva (September 10–12, 2003) was preceded by three Preparatory Committee meetings, which were all held in Geneva. PrepCom 1 lasted from July 1 to July 5, 2002, while PrepCom 2 was held on February 17–28, 2003. PrepCom 3 was preceded by an Intersessional Meeting that was held in Paris on September 15–26, 2003, and finally convened three times: PrepCom 3, September 15–26, 2003; PrepCom 3A, November 10–14, 2003; PrepCom 3B, December 5–6 and 9, 2003. The realization of the Intersessional Meeting and the threefold articulation of the PrepCom 3 became necessary because after the first two preparatory meetings multistakeholder interaction appeared to be more complicated than expected and the lack of clear procedural lines was almost jeopardizing the possibility of reaching a minimum agreement between the parts. Together with the PrepCom meetings, before Geneva five regional meeting were convened: the African Meeting (Bamako, May 25–30, 2002); the Pan European Meeting (Bucharest, November 7–9, 2002); the Asian-Pacific Meeting (Tokyo, January 13–15, 2005); the Latin America and Caribbean Meeting (Bàvaro, January 29–31, 2003); and the Western Asian Meeting (Beirut, February 4–6, 2003). As far as the WSIS2 is concerned (Tunis, November 16–18, 2005), again three Preparatory Committee meetings were realized. PrepCom 1 was held in Hammamet on June 24–26, 2004. Given the hostile environment for negotiations, PrepCom 2 and PrepCom

3 were moved back to Geneva and were held, respectively, on February 17–25, 2005, and September 19–30, 2005. A second round of PrepCom 3 meetings was held in Tunis on November 13–15, 2005. In this case, only four regional meetings were realized between 2004 and 2005: the African Meeting (Accra, February 2–4, 2005); the Asian-Pacific Meeting (Tehran, May 13 to June 2, 2005); the Latin America and Caribbean Meeting (Rio de Janeiro, June 8–10, 2005); and the Western Asian Meeting (Damascus, November 22–23, 2004). Two more subregional meetings were convened: the Second Bishkek-Moscow Regional Conference on the Information Society, November 16–18, 2004, in Bishkek (Kyrgyzstan); and the Pan Arab Conference on WSIS-Phase II, An Arab Regional Dialogue, May 8–10, 2005, in Cairo (Egypt).

13. Procedure-focused discussions were particularly lively in preparation for the first WSIS phase, while preoccupations were watered down before the Tunis phase (although they were not absent, in particular with reference to the design of the two specific bodies, the TFFM and the WGIG).

14. Beside the abovementioned articulations in working groups and caucuses, which were intended to channel the specific competencies of participants, during PrepCom 1 the Civil Society Plenary was officially recognized by civil society participants as the general decisional body for the whole group (Ó Siochrú 2004b). Furthermore, to smooth processes on contents and procedural issues, two more bodies were created: the Content and Theme Group and the Civil Society Bureau, the former in charge of gathering inputs coming from thematic groups and the latter in charge of dealing with governmental and private sector groups on the procedural level.

15. It might be worthwhile to recall here that multistakeholder processes are not aimed at subverting the functions of actors in the political process, nor they are aimed at collective drafting of provisions. Rather, they are aimed at gathering all relevant points of view and perspectives that should, at least in theory, interact with one another on an equal footing so as to generate informed (although not always consensual) results.

16. We refer here to initiatives such as the project Voice 21 or the People's Communication charter. See http://comunica.org/v21/ and http://www.pccharter.net.

Chapter 2

Investigating Content and Process in Political Dynamics

Theoretical Background and Analytic Framework

At the end of the first chapter we outlined several challenging points for the study of the broad Global Communication Governance (GCG) field, thus stressing in particular the need for more complete and articulated approaches to investigation that go beyond the perspective based on the concept of multistakeholderism. We argued indeed that the principal shortcomings characterizing existing analyses are the tendency to understand diversity only with specific reference to actors' attributes and to evaluate the scope of multiactor processes mainly with reference to impacts on official political outputs. While they look at the effects that formal and informal constraints to action have on the participation of different actor categories, existing studies on the governance of global communication do not pay systematic attention either to the relations actors establish among themselves or to how the agenda evolves consequently. However, in a context like the Internet Governance Forum (IGF), where no binding outcomes are foreseen and the collective construction of an Internet governance discourse comes to summarize the whole political interaction, overlooking the relational factor penalizes our understanding of multiactor dynamics and hides the consequences of collaboration between institutional and non-institutional actors. We therefore made a plea for an innovative and relation-based approach of study that looks simultaneously at the contents and processes of current debates on Internet governance (IG).

To respond to this plea, we start from a twofold assumption. First, we move from the idea that social construction of problems and *framing activities* through which actors attribute to particular situations the status of "problem" are crucial for the conduct of politics because they connect to actors' identities and link to a set of normative assumptions on what behaviors should be adopted under determined circumstances. In this sense, frames ground important decisions such as the type of action that should be undertaken to defend interests, the type of resources that should be mobilized, the partners that should be reached out to, and so forth. In a context where issue boundaries are so fuzzy, like the IG one, *framing activities* can produce highly conflictual dynamics, wherein different actors engage in order to make their vision prevail over that of others. Second, moving from the critique we made of the explanatory power of the multistakeholder perspective, we argue that considering exclusively the plurality of actors that are present on the political scene is not sufficient for grasping the relevance and the potential of multiactor processes, especially those that are taking place in the IG domain. Indeed, if we concentrate only at a formal level on the presence or absence of actors in the discussion, we can hardly understand the consequence of multiactor interactions in terms of collective construction and deployment of global political issues (and of IG in particular). We therefore propose to adopt a relational perspective according to which the relations established between actors in the IG domain constitute our main object of interest together with ties existing between the themes they carry into the discussion.

In this chapter, we illustrate the theoretical premises and the actual framework we employed in the study of the IG domain. We begin the theoretical journey that leads to the illustration of our analytic scheme with a closer examination of the underlying mechanisms that sustain the social construction of problems, thus emphasizing the role played by frames as those cognitive categories conveying particular meaning to situations and grounding actors' decisions to mobilize in a field. Despite the popularity of the concept of *frame* (e.g., Goffman 1974), there is some disagreement on how these cognitive categories tie to actual behaviors, especially in the political realm. To overcome this disagreement, we build on existing works on frames and collective action to isolate the main elements of a cognitive-behavioral chain that sustains political confrontations. Afterwards, we will illustrate the premises of a relational approach to the study of society and politics so as to offer an alternative way of looking at the multiactor interactions that develop in response to the thematic and procedural uncertainty brought by globalization processes. In considering networks as the main tool for translating such a relational view of society into practical research activities, we acknowledge its spreading use for the study of interdependencies established between institutional and non-institutional actors in contemporary political dynamics. However, this variety of uses generates a sort of conceptual confusion that, in the end, affects the heuristic potential of networks for the understanding of multiactor politics. Therefore, we will try to overcome confusion through a systematic review of four existing conceptualizations of "networks in politics" that, according to us, are particularly useful for exploring the political dynamics pertaining to the governance of global communications: *government networks*; *policy networks*; *collective action networks*; and *governance networks*.

We organize these four concepts within the umbrella of the *communication network* idea (Monge and Contractor 2003) with the aim of distinguishing between different ideal typical situations of interdependency between actors. This systematization effort originates from our approach to the study of global governance experiments (see Chapter 1), according to which the political globalization argument on political actors' networks of collaboration (or conflict) should be better specified by looking at the *causes*, the *actors,* and the *relations* that are established in the different cases. For example, multiactor networks generated to sustain traditional policy-steering activities will be different from networks of collective actions that rise, for example, in order to make governments responsible in social, political, economic, or environmental matters. While these two network types are both possible forms of political action in the global context, they respond to different needs, involve different actors, and are sustained by different relations. Therefore, the systematization we offer in this chapter is meant to associate the different ideas of "networks in politics" to a specific (although ideal typical) situation so as to limit conceptual overlaps and blurred explanations.

Building on this theoretical overview, we will then illustrate the analytic framework we employ for the study of IG dynamics stimulated by the realization of the IGF. We begin from an illustration of the concept of *discourse* that, according to our approach, comes to summarize the bulk of political activities in the IG domain as it is "metaphorically extended from its original roots in interpersonal conversation to the social dialogue which takes place through and across societal institutions, among individuals as well as groups and . . . political institutions themselves" (Donati 1992:138). We locate discourse production within networks of interaction between actors, and we apply this perspective to the resolution of the twofold uncertainty on contents and processes characterizing the IG case. Thus, in order to study this peculiar intertwinement between contents and processes, we use both semantic and social networks, where the former are structures whose exploration helps in uncovering the patterns of collective construction of the IG agenda and the latter represent political relations that develop between all different actors participating in the definition of the IG discourse. We propose to trace semantic and social networks established both online and offline as we consider both spaces as relevant arenas for the collective construction of an IG discourse.

1. STRUGGLING OVER MEANINGS: SOCIAL CONSTRUCTION OF PROBLEMS, FRAMES, AND FRAME DISPUTES

In general, problems do not exist as absolute conditions: their existence depends on social actors' perception and definitions (Blumer 1971). In this sense, as Hilgartner and Bosk pointed out, problems are "projections of collective sentiments rather than simple mirrors of objective conditions in society . . . [they are a] *putative condition* or situation that is labeled a problem in the arenas of public discourse and action" (1988:55, emphasis added). For example, during the NWICO debate the non-aligned countries perceived the unbalanced situation of world information and communication fluxes as a relevant problem. Consequently, they decided to take action to bring the situation back to "normality" and work for the establishment of a more democratic and balanced communication world order. In the eyes of the Western bloc, instead, it was the free flow of information that constituted "normality," while the Non-Aligned Movement (NAM) reaction was perceived as problematic. Consequently, the Western bloc decided to oppose the call for a NWICO, and the United States and the United Kingdom withdrew from UNESCO.

To understand why some actors label a situation a problem while others do not, it is important to remember that "people approach the world not as naïve, blank-slate receptacles who take in stimuli as they exist in some independent or objective way, but rather as experienced and sophisticated veterans of perception" (Tannen 1993:20). Individuals and groups decode life experiences and novelties using specific cognitive categories that are both derived from past experience and socially learned. The well-known concept of *frame*, understood as the "schemata of interpretation that enables individuals to locate, perceive, identify and label" reality and its realizations (Goffman 1974:21), is often employed to describe the act of organizing experience cognitively and taking action consistent with the expectations connected to interpretation. Through frames, individuals and groups recognize, give meaning, and manage all the concrete situations they find themselves in. When crossing a certain event on their path, individuals apply what Tannen calls *interactive frames,* which refer to a "definition of what is going on in interaction without which no utterance (or movement or gesture) can be interpreted" (1993:334; see also Tannen and Wallat 1999). Once a situation is decoded, *knowledge schemas* are activated, and so are "participants' expectations about people, objects, events and settings in the world" (Tannen 1999:334), from which a particular behavior will follow. Conveying a certain meaning to a situation, then, is the first step toward action.

However, despite the widespread adoption of the concept of *frame* for describing the processes through which we recognize the world around us, there is also a widespread disagreement on the way they link to actual courses of action (Fisher 1997). Indeed, the link between cognitive resources, meaning attribution, expectations, and behaviors has been conceptualized in different ways depending on the theoretical tradition and the final aim of interaction (for a detailed review, see Fisher 1997). This sort of disagreement also permeates the study of political dynamics, where, from international relations studies (Adler 1997; Onuf 1989; Wendt 1992, 1999) to collective action research (Benford and Snow 2000; Johnston 2002; Snow and Benford 1988, 1992), frames have been largely employed to explain actors' behaviors starting from the assumption that actual action courses follow the interpretation of the world.

In spite of the different conceptualizations that have been provided, we can extract from existing literature on the relevance of frames for politics four main elements that help us understand the chain that goes from cognition to action:

- frames rely on cognitive resources that are internal to the individual and serve the purpose of organizing the world;
- events and discourses are organized on the basis of selected frames: what is not recognized as belonging to a frame is left outside, perceived as irrelevant, ignored, or, in certain cases, rejected as "inconceivable";
- frames are not mutually exclusive; experience can (mis)match with more than one frame;
- the content of frames is both fixed in a certain point in time and also constantly modified through social processes of interaction.

It should be pointed out that there is a neat preference in literature for the noun *frame* rather than the verb *framing* (see Johnston 2002), and this has yielded a rather static idea of frames as synonymous with situational meaning, thus paying not enough attention to how they can vary over time. In fact, both the synchronic and the diachronic dimension of frames imbue political processes: political confrontation (and conflict) can generate from the acknowledgment of frame inconsistencies (i.e., some of them frame a certain circumstance as problematic whereas some others do not) or can be justified by the attempt to modify the way other actors frame a certain issue.

Indeed, every policy issue has its own "culture"; that is to say, there is an overall conceptualization of a particular topic that is composed of a multiplicity of elements—metaphors, symbols, values, expectations, and the like— that translate into a public discourse (Gamson 1988:221–222; Gamson and Modigliani 1989:1–2). Within a certain issue culture there can be a multiplicity of different frames, or *interpretative packages* (Gamson and Modigliani

1989), that combine some of these elements in a different fashion. As a consequence of the activation of specific frames to interpret reality, some political actors will be concerned with certain situations whereas it is very likely that their adversaries will not have the same perception because they have applied different combinations of cognitive elements to interpret the same piece of reality. To clarify this mechanism using a common example, we can think of nuclear energy, which can be depicted as a tremendous economic and technical advancement or, conversely, which can be seen as a dangerous and risky mode of energy production. We could expect that parliamentary groups of opposed orientations would differ in the way in which they portrait nuclear power, its environmental consequences, and, more broadly, even progress. The underlying cognitive divergence that separates the two groups is very likely to generate a dynamic of opposition that will be played out at different levels (e.g., supporting sensitization campaigns, fighting over policy provisions for the energy sector, etc.). However, positions held by arrays different sides are not immutable. For example, as a consequence of major disasters, those who promote nuclear energy can step back and reshape their frame toward a more cautious position that acknowledges the risks and, in the end, reduce the "cognitive distance" from political adversaries.

The dynamics of frame definition and evolution are played out at two different levels: policy frames or packages are constantly negotiated at what Gamson and Modigliani (1989) call the *cultural level,* and, almost simultaneously, they interact with individual perceptions and determine the appeal of political action in public opinion at the *individual level*. The higher the appeal, the higher the chances that political action will be perceived as legitimate and appropriate (Gamson 1988; Gamson and Modigliani 1989; Snow et al. 1986; Snow and Benford 1988), and the higher the political leverage.

At the "cultural level," competition between frames becomes even more important when political processes develop around controversial and new social problems whose threats or consequences are not well defined. Especially when the boundaries of an issue are not neat, the relevance of the symbolic elements and of cognitive resources is high and "discourse becomes an explicit battle-ground for ideological wars of position that are dynamic products of dialogic interaction" (Steinberg 1998: 853). Depicting one issue as problematic or not is what grounds the dichotomy between change and conservation of the status quo, thus constituting the first step toward mobilization and resource activation. When the struggle is focused on how actors portray a certain issue, the cultural feature of this "symbolic challenge" (Melucci 1996:182) by no means implies that the confrontation is less severe or radical than when conflict is experienced through physical contention. In the first place, a symbolic challenge can always result in physical and violent manifestations. But even when they remain at the conceptual level, symbolic challenges can have harsh consequences, as happened in the NWICO case when different ways of interpreting communication processes led the United States and the United Kingdom to their withdrawal from UNESCO. The severity of battles fought over meanings is justified by the fact that what is at stake is the articulation of a dominant vision on the issue (a "master frame" [Snow and Benford 1992]) that will highly influence the policy issue culture and, hence, future policy provisions. Indeed, *frame disputes* (Benford 1993) can develop in relation to different aspects: the cognitive extension of the frame (*diagnostic frame disputes*), that is, which part of reality is organized within a certain frame; the way in which problematic aspects of reality should be managed (*prognostic frame disputes*), that is, the possible behavioral patterns that are accepted for acting upon a certain situation; and how reality should be presented in order to maximize mobilization potential (*frame resonance disputes*), that is, what strategies will gain the support of a vast portion of public opinion. In sum, prevailing over other competitors in frame disputes generates political leverage because it allows a competitor to impose a certain vision of reality, of what is right and wrong, of what should be done now and in the future.

However, as Benford (1993) points out when speaking of *frame resonance*, prevailing over other competitors is not enough. Political actors have to maintain a sort of "cognitive tuning" with a larger audience to ensure the public support of their political initiatives (Snow and Benford 1988). In this context, strategic alignment processes between frames at the individual and at the cultural level (Snow et al. 1986) are extremely relevant for actors to reach and maintain legitimization (Chilton 2004). When dealing with a specific issue, political actors do engage in the formation of a policy discourse, and those who activate frames that resonate louder with public opinion are in a privileged position. It should be noted that the strategic alignment of frames between the cultural and the individual levels is, as the label suggests, a *strategic* activity. This implies that this mechanism does not take place as a natural consequence of a mutual correspondence between political actors' views and citizens' needs. More often, to avoid significant power losses, political actors deliberately reframe their positions in a way that guarantees them public support for their strategies. In other words, the strategic element of alignment only seldom corresponds to a representative or democratic use of frames but is, nonetheless, one of the keys for political success.

One example of strategic alignment in the IG domain is provided by the passage from a focus on cyberdemocracy to cybersecurity after the 9/11 events (Kleinwächter 2004; see also Chapter 3). Those groups within institutional IG bodies who wanted to increase the levels of network security without raising concerns about excessive controls, censorship, or violations of net neutrality aligned their discourses with the spreading concern over global terrorism. Starting from tragic real-world events, some parts of the IG institutions reframed a discourse *on control* into a discourse *on security* and, in this way, overcame general skepticism and encountered the wide agreement of all those who were afraid of a terroristic use of the Internet (especially in the North of the world). After this alignment, the levels of control on the Internet could be increased and, as this was portrayed as a necessary measure to combat terrorist attacks, critiques on possible violations of individual privacy, for example, found less support among the public, which, instead, recognized more easily these measures in terms of strategies to ensure general safety.

2. STUDYING INTERDEPENDENCIES THROUGH NETWORKS

There is a *fil rouge* that joins together narratives on globalization, global governance, and multistakeholderism, namely, the accent on networked forms of collaboration or, more broadly, on interactions established between institutional and non-institutional entities, be these latter representatives of strong economic interests, non-governmental organizations (NGOs), informal platforms, or knowledge communities. This focus on interaction is justified by a "decentralized concept of social organization and governance [for which] society is no longer exclusively controlled by a central intelligence (e.g., the state); rather controlling devices are dispersed and intelligence is distributed among a multiplicity of action (or 'processing') units" (Kenis and Schneider 1991:26). However, to understand how such a distributed environment is structured and what consequences it has for the conduct of politics, it is not enough to acknowledge the presence of a plurality of actors on the scene. Rather, given the features of dynamism and complexity that characterize the global context (Kooiman 2003), new approaches are needed to investigate the plurality of actors *and* the interactions in which they engage, with the overall aim of uncovering the implications of multiactor collaborations (or, perhaps, conflict) and the possible redistribution of power among them (Padovani and Pavan 2011). To this end, networks are a powerful image for portraying the growing complexity of contemporary societies: they represent the principal feature of a new "social morphology" (Castells 1996, 2000) for which policy outcomes are "generated within multiple-actor-sets in which individual actors are interrelated in a more or less systematic way" (Kenis and Schneider 1991:32). Thus, not only do networks constitute a lens for depicting and reducing the complexity of situations, but their emergence is nowadays also considered a relevant political result, as it was stressed in the case of civil society participation in the World Summit on the Information Society (WSIS) process.

The usefulness and the pertinence of a relational approach to examining complex transformations innervating (not only) political processes can already be seen in the very origins of the structural approach to the study of society, in particular within Simmel's analysis of *the web of group affiliations* (1955 [1908]). Indeed, Simmel refused to consider individuals as atomized and isolated and conceived social ties as the very key element to understanding the diversity, dynamics, and complexity of modern societies. Focusing more closely on political dynamics, a structural approach built on Simmelian bases and centered on the study of relational patterns has been employed mainly in two directions: on the one hand, to examine the formation of networks and coalitions within states, with a specific accent on explanations for resource mobilization; and, on the other, to study the asymmetric relations between states that can result in dependency systems (Wellman 2002 [1988]).

And yet, especially in the study of politics, both the structural approach and the network concept continue to be applied diversely and not always systematically, to the point that metaphorical uses of networks keep paralleling the implementation of relational views into actual methods for analysis and research programs (Wellman 2002 [1988]; Diani 2003). The scarcity of empirical investigations that do not consider networks only as metaphors (Anheier and Katz 2004; Katz and Anheier 2006) aggravates the lack of integration between the "soft" and the "hard" sides of the network concepts and generates consequences at two levels. First, the basic assumptions of a relational perspective for the study of society and of political arrangements are often taken for granted while they should constitute the benchmarks for empirical investigations. As networks are simply taken as metaphors, the realization of the actual potential of the relational approach in comparison to more traditional attribute-based perspectives is jeopardized and the usefulness of networks as investigation tools ends up being questioned. Second, networks' specificities are often ignored, as if the mere acknowledgment that actors are joined together in a system of interdependency would be enough to grasp the consequences of transformations characterizing global politics (Rhodes 1997). Before we proceed to the illustration of how networks can be applied to the study of IG multiactor dynamics, both these points need further clarification.

2.1 A "Network" Point of View

According to Knoke and Kuklinski, networks are "a specific kind of relation linking a defined set of persons, objects or events" (1982:12). What does it imply to look at society starting from these premises? Marin and Wellman (2010) point out that the adoption of a network perspective implies first of all that the core elements under examination are *relations* linking actors together and the *patterns* in which these relations are combined. In this sense, the adoption of a network approach implies a whole different way of looking at society, one for which "the structure of relations among actors and the location of individual actors in the network have important behavioral, perceptual, and attitudinal consequences both for the individual units and for the system as a whole" (Knoke and Kuklinski 1982:13).

Going into more detail, Wellman (2002 [1988]) observes that the adoption of a network approach to the study of society implies five main assumptions:

1. A shift from interpretations of actors' behaviors that are based on attributes toward interpretations that are grounded on structural constraints to activity. This means that behaviors are not explained starting from who an actor *is* but rather starting from what she does or does not do within a certain environment. Individual characteristics help us understand why a certain action is undertaken, as we start from the assumption that individuals with similar attributes occupy similar positions within an interactional *milieu* and, therefore, might experiment with similar constraints to action.
2. A shift from the attempt to sort units into categories toward an analysis centered on relations between units. In other words, social actors should not be grouped on the basis of their attributes but, rather, on the basis of the relations they establish. Once again, individual characteristics do not lose their importance but contribute to the explanation of relation-building patterns.
3. A specific attention to influences that derive from relations existing between other actors in the network. This means that constraints and possibilities for action also derive from relations that are established in the interactional *milieu* within which the observed actor is immersed and not exclusively from his or her attributes and relations.
4. A specific attention to the overall context; that is, a structure is a "network of networks" that does not necessarily have to be partitioned into different groups.
5. The necessity of developing analytic methods that go beyond mainstream statistics and exploit the relational nature of the social structure.

Moreover, looking at society through networks means also focusing on a specific organizational form that tends to emerge under specific circumstances.

In this regard, the main tendency is to adopt a generalist conception of networks as loose forms of organization that develop as alternatives (or in opposition) to markets and hierarchies, while only seldom is it adequately stressed that networks tend to emerge under specific conditions to *incorporate, supply,* or *challenge* (not to substitute for) markets and hierarchies and, more generally, other forms of regulation (Kahler 2009). Going directly to the roots of networking processes, Powell (1990) argues that market, hierarchy, and network organizational forms can be distinguished starting from interactions developed among actors within a particular system. He suggests that network arrangements are preferred when the quality of items exchanged between actors cannot be measured univocally and relations linking actors together are long-term but are not developed within a binding or legal framework. In other words, when actors must sustain their interaction over time but there is no one central authority (a state or an industry) that is able to set a common standard or to impose effective sanctions against deviant behaviors, networks arise to reduce the overall uncertainty that derives from such a lack. For networks are based on cooperation, foster mutual learning and the spread of knowledge, allow for a fast translation of knowledge into action, are flexible enough to compensate for the variability and uncertainty characterizing resources, and foster the employment of knowledge and technical innovation (Powell 1990:322). Similar considerations are developed by Jones, Hesterly, and Borgatti (1997), who observe that network arrangements are adopted when relations between parties must be sustained over time within a context where there is high uncertainty regarding future developments, where changes cannot be predicted in advance, and where actors have different skills and capacities and goals to be pursued.

The fact that networks are organizational modes that allow for an easier management of uncertainty is also the starting point for considering them as very useful tools in the study of political arrangements that follow the retrenchment of the nation-state as the sole locus of authority. In his well-known analysis of changes in British politics, Rhodes (1997) argues that the traditional monocentric or unitary government has been replaced by a multicentric and multilevel governance system in which institutional agencies are not only linked to one another across levels but are also integrated, supported, and limited by a plurality of non-institutional entities (e.g., private companies) that fully participate in the determination of policy outcomes. The author sees in *policy networks* the appropriate way of organizing the "multiform maze of institutions that makes up the differentiated polity" (Rhodes 1997:3) to revitalize steering tasks through the collaboration of institutional and non-institutional actors. In other words, in a globalized political context, the emergence of networks "is a result of the dominance of organized actors in policy making, the overcrowded participation, the fragmentation of the state, the blurring boundaries between the public and the private, etc." (Kenis and Schneider 1991:41). Networks are preferred to markets and hierarchies as modes of organizing political processes with specific reference to three aspects: *the relations established between actors,* which are pluricentic as opposed to monocentric, entailed by state regulation and multicentric arrangements characterizing market competition; *the decisional mechanisms enacted,* which are based on reflexive rationality rather than on the substantial rationality characterizing state regulation or on the procedural rationality defined by markets; and, finally, *the level of compliance with collectively negotiated decisions,* which is ensured not by the means of coercion typical of the state or by the *"fear of economic loss"* but rather through the generation of trust and political obligation (Sørensen and Torfing 2007:11–12). In sum, network arrangements are adopted because steering activities about complex matters require the simultaneous presence of diverse actors and competencies: it is along network ties that participants' points of view can be coordinated and consensus is possibly achieved (Börzel 1997).

And yet, despite their widespread adoption in the investigation of emergent governance arrangements, there is an overall lack of consensus on what networks really constitute for the study of politics: a mere metaphor, a method, an analytical tool, or a proper theory (Börzel 1998). In this context, the predominance of a generalist use of networks as synonymous for loose organizational structure hides some of the most relevant characteristics of networked politics because it conveys an artificial idea of powerless horizontality and assumes that network dynamics are aimed only at consensus building (Padovani and Pavan 2011). For example, the analysis of how civil society at the WSIS arrived to draft collectively its final documents took advantage of a metaphorical use of networks to stress the interconnectedness and the mixing up of perspectives that characterized the process. However, a more systematic analysis based on networks showed that the positional differences of civil society representatives within the interactional milieu created by public interest organizations during the WSIS actually translated into different capacities to influence the final document and, therefore, implied a different political leverage (Mueller, Kuerbis, and Pagé 2007).

Coming to the very object of our attention, the IG domain, there are several aspects that indicate the appropriateness of networks as tools for the analysis of multiactor dynamics fostered by the IGF. In the first place, communication issues are matters whose qualities are hardly measurable or, more precisely, whose qualities are differently measured according to different actors. Business entities can adopt prices as their point of reference, whereas governments, for example, might choose to base their evaluations on the effectiveness of

security enforcement mechanisms. Since both governments and private entities are all interested and legitimate claimants in the game, their perspectives
and needs have to be coordinated. In this context, networks provide a way to
perform this complicated task. Secondly, if it holds true that institutions are
generated through long-term repeated interaction (Jepperson 2000; Wellman
2002 [1988]), then the redefinition of an IG discourse should be read in the
first place as starting from the links established between actors thanks to which
thematic boundaries and interactional patterns are being redefined. Moreover,
in a multistakeholder institutional context (such as the WSIS or the IGF)
where large opportunities for constructing political relations are provided, it
cannot be assumed that actors grouped under broad categories such as civil
society, businesses, or institutions will behave homogeneously either in their
seizing of opportunities or in the behavioral patterns they will actually follow.
Hence, instead of looking exclusively at who is participating and who is not,
we should study how those who are participating relate to one another and how
the absence of other actors influences the overall structure of relations.

2.2 Interdependency in Action: More than One Idea

It is not only the metaphorical use of networks that jeopardizes their potential
for disentangling the complexity of contemporary political arrangements. In
fact, to describe the myriad concrete dependency relations that characterize
an environment of "distributed intelligence" (Kenis and Schneider 1991)
like the one we live in, the concept of networks has been reelaborated and
customized so many times that we easily get lost in conceptual confusion.
Similar situations are labeled differently, the same label is applied to different occurrences, and the underlying assumptions leading to the adoption of a
specific network concept over all the others are seldom made explicit (Börzel
1998). In this context, the absence of an overall systematization effort where
each network concept corresponds to a particular type of political interaction
constitutes a major hindrance to the systematic study of networked politics
because it reduces the possibility of examining contemporary complexity
by looking at the specific characteristics of established networks and at the
type of interaction that takes place within them.[1] On the contrary, we believe
that, in order to better understand contemporary governance processes in
the information and communication field, it is suitable to better specify the
"political globalization argument" on the networks formed by political actors
by looking at the *causes,* the *actors,* and the *relations* that are established in
the different cases (see Chapter 1).

    In order to remedy this lack, we do not need to invent new concepts or
labels. Rather, we can start from existing applications of the network idea to

the study of political transformations and try to unify them under an overarching perspective. Four conceptualizations of "networks in politics" seem particularly suitable for becoming part of this conceptual clarification exercise:

- *Government networks* (Slaughter 2004)
- *Policy networks* (Adam and Kriesi 2007; Börzel 1998; Hanf and O'Toole 1992; Kenis and Schneider 1991; Knoke et al. 1996; Laumann and Hintz 1991; Laumann and Knoke 1987; Marin and Mayntz 1991; Marsh and Smith 2000; Rhodes 1997; van Waarden 1992)
- *Collective action networks* (Diani 2008, 2009; Diani and Bison 2004)
- *Governance networks* (Sørensen and Torfing 2007)

Although these conceptualizations were developed separately and in different historical moments, they all share two main characteristics. First, they all stem from the assumption that globalization processes influence political dynamics, stimulating an increased level of interaction and interdependency between governmental and nongovernmental actors. Second, they are associated with a multiplicity of meanings, depending on their authors and the scientific purpose they serve. In this regard, if conceptual clarifications have been attempted, they were pursued in relation to one single concept (see, for example, the review of *policy networks* done by Börzel [1998]), whereas a "relational review" that unifies them under the same interpretative perspective has not been pursued yet.

What are the advantages of a relational review of these four network ideas? In the first place, if we join together these four conceptualizations and understand them as a specific and ideal typical form of networked politics, each label is associated with a more clear-cut meaning and describes a particular type of interaction that is not addressed by the other three. In this way, meaning overlaps are minimized and, consequently, the usefulness of each concept for the analysis of different dynamics in the IG domain (or other GCG domains) is more evident. Second, in order to choose the most appropriate network form to describe the actual process we are observing, we are forced to start necessarily from the type of interaction we are studying and to identify the causes for which networks arise, the actors that belong to them, and the relations they engage in. In this way, we aim at fostering the deliberate and cautious adoption of networks as interpretative tools, thus avoiding generalist and vague uses, as the choice between one label over another depends on a characterization of the type of interdependency that we are looking at.

How can we hold together all these heterogeneous ideas? The starting point to accomplish our task is provided by the fact that all these forms of network point to the multifaceted deficit that states are affected by within global settings (Hockings 2006; see also Chapter 1). Because they are not self-sufficient, states must collaborate with other actors, but this collaboration creates an overall situation of *interdependency* among institutional and non-institutional entities. Interdependency is played out, first and foremost, through the construction of communication relationships between those actors who are in need for and those who possess the knowledge and the resources that are necessary to manage complex tasks (Knoke 1990:11). Consequently, *government networks, policy networks, collective action networks,* and *governance networks* can all be seen as different forms of *communication networks* that join together actors mainly through the exchange of messages across time and space in the attempt to stabilize structures of interaction out of the chaos provided by the globalized context (Monge and Contractor 2003:11–18).

Starting from this premise, we can distinguish between different forms of communication networks, keeping in mind that (a) networks are determined by *actors* (nodes) and *relations* (ties); and (b) the focus here is on governance experiments, that is to say, that networks are set up in order to achieve some final goals or purpose. Overall, then, *government networks, policy networks, collective action networks,* and *governance networks* can be considered different forms of communication networks that can be differentiated on the basis of three elements:

1. *Actors.* Exchanges are established among actors of the same nature (*within* groups, such as between governments or between NGOs) or among actors of different natures (*between* groups, for example, between governments and industries).
2. *Relations.* Interactions can involve the exchange of different kinds of resources, both material (e.g., money) and non-material (e.g., information).
3. *Purpose.* The condition of interdependency can be identified with reference to different types of "public purpose" (Sørensen and Torfing 2007): it can affect traditional policy-making processes where steering activities are involved, or it can be played out within dynamics that are aimed at generating ideas and common understandings or at posing the bases for the emergence of norms.

Far from being in contradiction with the premises of the structural approach outlined above, the prominence given here to *actors' attributes* (i.e., their nature as institutional or non-institutional entities) is helpful in studying multiactor dynamics in contemporary global arrangements as well as in compensating for the overall inattention to interdependency relations established among relatively homogeneous sets of actors (especially of an institutional nature). Indeed, changes in the organizational modes adopted by homogeneous sets of actors are rarely depicted in terms of innovation, or they are considered as relevant as the study of processes of enmeshment between state and society. Attention is mainly focused on those forms of collaboration between state and non-state actors, whereas, if governments coordinate among themselves in a new way or if public interest entities organize transnationally, these processes are not studied with the same curiosity or they stimulate the production of ad hoc frameworks of analysis. As far as *relations* are concerned, it is necessary to specify that sets of actors are joined by one relation at the time, whether this is simple (e.g., "giving money to") or composite (e.g. "giving money to" and "trading"). Therefore, looking at different relations means looking at different networks.[2] Finally, the explicit attention paid to the *purpose* of networking is an acknowledgment of advancements made in theorizing networks as governance tools. Indeed, as pointed out by Sørensen and Torfing (2007), we recently moved from a first generation of network studies (mainly preoccupied with justifying the existence of networks as different organizational modes for policy steering) to a second generation that focuses on how networks operate and on the results they can achieve. In this switch, the final outcomes of networked governance can be depicted in terms of *public purpose,* which is "an expression of vision, values, plans, policies and regulations that are valid for and directed towards the general public" (Sørensen and Torfing 2007:10). In this sense, attention should be paid not only to networks producing policy outputs but also to those leading to the achievement of informal policy outcomes such as common frames, norms, and shared understandings.

If we combine *actors, relations,* and *purpose,* we can distinguish networks in a twofold way. Recalling the need to explore global governance experiments by looking at causes, actors, and interactions (see Chapter 1, section entitled "Global Governance"), we can separate types of communication networks first of all on the basis of what *actors* are involved and the reason why they are in networks (i.e., the network *purpose*). Moreover, further distinctions between each network type we identify in this way can be made if we look also at the type of interaction actors engage in, that is, if we look at the *relations* established. For example, among networks set up by homogeneous actors for generating a shared policy provision (first level of distinction), it will be possible to trace and study a network for every resource exchanged (information, money, collaborators, trust, mistrust, etc.). Characterizing the network type, as we mentioned above, allows us to distinguish between the different types of interdependencies we are studying. The study of network subtypes based on the relations established by actors helps us instead to deepen our knowledge of specific processes that are deploying in a certain area.

Therefore, if we stop for now at the first, more general level, we can distinguish *government networks*, *policy networks*, *collective networks,* and *governance networks* as forms of communication networks on the basis of the following:

1. *The levels of morphological diversity characterizing the actors involved.* This dimension opposes communication network forms that develop within sets of homogeneous actors to those that can be developed by sets of heterogeneous actors.[3]
2. *The purpose of interaction.* This dimension opposes communication network forms that develop to sustain conventional policy-making processes to those aimed at fostering the production of norms and common frames.

Crossing these two dimensions, we obtain a systematization like the one depicted in Table 2.1, for which

1. *government networks* describe joined activities pursued by governments and intergovernmental organizations with the aim of reaching consensus on rules and practices to be developed and implemented within an interconnected and multilateral world system;
2. *policy networks* describe situations in which conventional policy making is sustained by exchanges between state actors and non-institutional actors;
3. *collective action networks* represent those forms of interaction between non-governmental groups in the attempt to impact political processes with expertise and alternative and informed points of view; and
4. *governance networks* represent those forms of interaction that are aimed mainly at creating shared norms, perspectives, and frames among a plurality of actors of different natures.

In this systematization, each network concept comes to represent an ideal typical situation and is associated with a precise type of dynamic of interaction between certain set(s) of actors.[4] There is no neat separation line in the table, so to suggest that different network types can coexist and can link to one another within the same domain. As suggested above, for each of the communication network forms in our scheme, different subtypes can be identified, starting from the relational content that is joining together actors. Indeed, many different strategies can be pursued in order to achieve the final purposes that join together actors' set(s).

The IG domain, like all other GCG domains, is inhabited through all these types of networks. What type of communication network describes better the

Table 2.1.   Framework for the classification of communication networks

		Purpose of Interaction	
		Policy Making Oriented	Cultural/Ideational Outcomes Oriented
Levels of Morphological Diversity	Lower	Government networks	Collective action networks
	Higher	Policy networks	Governance networks

collective construction of an IG discourse stimulated by the IGF realization? We mentioned already several times that the IGF was created to provide a space for multistakeholder dialogue and has no formal commitment for the production of binding outcomes. However, we also argued that the absence of conventional policy-making dynamics does not lower the political relevance of the process itself. If we follow the classification scheme just illustrated, given the heterogeneous nature of the actors involved and considering that the purpose of the interaction is to collectively redefine the boundaries of the IG domain and the roles played by different constituencies within it, then the networks we are analyzing are *governance networks*. However, it remains to be assessed how we empirically investigate them. To do so, we need to move from conceptual clarification efforts to a more proper analytic framework that will guide our analysis.

3. A FRAMEWORK FOR THE ANALYSIS OF GLOBAL COMMUNICATION GOVERNANCE DYNAMICS

In the previous sections we saw how relevant frames and networks can be for the analysis of political processes in the global era, and we provided some insights on their usefulness for inquiring into current IG debates. Also, we saw that the concept of governance networks, understood as those forms of interaction that are aimed mainly at creating shared norms, perspectives, and frames among a plurality of actors of different natures, should be the preferred lens through which we investigate multiactor dynamics fostered by the IGF. However, so far we have explored frames and networks separately, and we still need to conceptually join them together. Moreover, we need to reflect critically on what spaces for their development we should explore. An extended conceptualization of *discourse* provides us with the starting point for pursuing our first task, that is, conceptually joining political contents and processes. Thus, we will argue that the development of an IG discourse is being carried on both offline, thanks to occasions such as the meetings of the IGF, and online, thanks to the multiplication of virtual spaces of discussion that carry on multiactor interaction in between the official meetings.

3.1 Grounding Discourses in Network Ties

If communication provides the overall linkage between actors within the IG domain and, more broadly, in the GCG field, then actual exchanges between them can be read in terms of their contribution to the collective construction of a *discourse*. Here, discourse must be understood as "metaphorically extended from its original roots in interpersonal conversation to the social dialogue which takes place through and across societal institutions, among individuals as well as groups and . . . political institutions themselves" (Donati 1992:138). Looking more precisely to the IG domain, as the IGF has not been committed to any steering function but is only meant to provide a space for multistakeholder policy dialogue, the construction of the discourse comes to summarize the essence of political activity as it entails the production of a set of "concepts, categories, ideas, that provide its adherents with a framework for making sense of situations, embodying judgments and fostering capabilities" (Dryzek 2005:1). Going back to the concept of policy domains (Laumann and Knoke 1987; Knoke et al. 1996; Pappi and Knoke 1991), the more discourses crystallize the more the contours of a policy domain will be neat and, consequently, access levels to it will be narrower. Shaping discourses implies, at the same time, the redefinition of the thematic boundaries of a domain and the progressive consolidation of interactional patterns through which all participants deliver their contributions. Hence, *discursive practices* lower conceptual and procedural uncertainties, thus identifying mechanisms for interests' intermediations.

The political value of discourse deployment goes then beyond the fact that it might or might not culminate with the production of binding political outputs. If we adopt Melucci's definition of political relations as those "which are activated in order to reduce uncertainty and mediate among conflicting interests through decisions" (1996: 211), then it becomes evident that multiactor interactions developed within the IGF space are *inherently political* despite the lack of any commitment to the production of binding provisions. In this specific case, decisions will pertain to IG contents as well as to the roles and the responsibilities of governments, business entities, and civil society groups in the IG domain. Thus, these decisions will not be codified through binding provisions but crystallized in the institutionalization of political behaviors that will emerge from experiments on the ground in the formation of an IG discourse. Here lies the very potential of looking at IG dynamics from a discourse point of view: under certain circumstances, indeed, discourses translate into normative frameworks binding political behaviors. According to Khagram, Riker, and Sikkink (2002), this happens once issues are framed through communicative exchanges; then are confronted and further articulated with reference to more consolidated issues and norms; and, finally, put into the agenda and articulated into statements.

In general, then, discourse consolidation precedes the emergence of norms or, in other terms, of "shared expectations of standard behaviors for actors with a given identity" (Finnemore and Sikkink 1998:892). When these normative assumptions are generally accepted, they constitute the base for the institutionalization of political roles and functions (Jepperson 2000). In this context, it is evident that "the power to shape the agenda or to shape the very manner in which issues are perceived and debated can be a substantial exercise of power" (Sikkink 2002:303–304). Hence, looking at how discourses develop as well as at the actual contributions delivered by actors involved is far from being a secondary task in the analysis of global political transformations. Thus it becomes even more relevant when the possibility of officially contributing is provided also to non-institutional entities, as in the IGF case. In contexts like these, discourse becomes a battleground (Steinberg 1998) where confrontation is deployed, at the same time, in relation to thematic and procedural aspects of political dynamics.

Discourse is produced within networks that are shaping and, at the same time, shaped by cultural and communicative interaction between actors themselves (Mische 2003:258). From an analytical perspective, if we consider that the collective construction of an Internet governance discourse is aimed at lowering the IG twofold uncertainty and is taking place within governance networks fostered by the IGF process, then discursive dynamics should be investigated by tracing the following:

- *Semantic networks*, which "map similarities amongst individuals' interpretations" (Monge and Contractor 2003:173). Tracing and analyzing semantic networks contributes to the investigation of how thematic uncertainty is reduced, focusing not only on what themes have entered the discourse but, more relevantly, on how the presence of a multiplicity of actors affects the aggregation of themes in discursive threads or the creation of disconnectedness patterns among issues.
- *Social networks*, which consist of a "finite set or sets of actors and the relation or relations defined on them" (Wasserman and Faust 1994:20). Actors, in this context, are "discrete social entities" of different types: individual, corporate, or collective (Wasserman and Faust 1994:17). Tracing and analyzing social networks contributes to the investigation of how procedural

uncertainty connected to the proliferation of actors on the political scene is managed interactively through the establishment of relations but also through the exclusion of some actors from interaction.

At both levels, frames and connections are acting together, and, indeed, semantic and social networks should not be conceived as alternatives to one another but rather as complementary. However, when we look at the former, our attention is focused on the very construction of an IG agenda, whereas, when looking at the latter, the analysis aims at uncovering the underlying logics that guide partnership-building strategies in a multiactor environment. It is evident that there is a strict link between the two types of networks: depending on agendas, certain sets of actors will be allowed to enter the discussion and, in turn, their interaction will contribute to further refine the agenda, thus influencing actors' opportunities to enter and act within the political domain.

3.2 The Importance of Being Online

There is a second dimension beside the social or semantic one that we need to consider in drawing the premises of our analytic framework. So far, we have organized our theoretical journey by focusing mainly on the real, offline world, while we have seldom mentioned that networks, interactions, interdependencies, and frames can be developed also in the virtual space opened by Information and Communication Technologies (ICTs) and by the Internet in particular. It is then necessary to elaborate more closely on the value of the virtual space as an environment for the development of political dynamics. A deeper look into this aspect is even more relevant in the case of IG, where the Internet not only creates a further political space that is less affected by the limitations characterizing offline processes (Padovani and Pavan 2007, 2008) but is also the very matter of contention.

The redefinition of the nation states' role in the global context brings with it also the modification of the public sphere, which cannot be conceived any longer as linked to the Westphalian state order (Habermas 1989). Fraser (2005) points out how contemporary transformations generate a transnational public sphere that is multilayered and multilevel exactly like the sovereignty rearrangement it corresponds to. In this context, ICTs and the Internet in particular play a major role because of the vast range of communication possibilities they offer beyond time and space differences. Accusations of technological determinism were moved against arguments depicting the Internet as a democracy medium, but these do not consider that all mass media have been enthusiastically depicted as tools for democracy when they first appeared on the scene (Calhoun 1998; Polat 2005). Such enthusiasm is due to the fact that media spread the relational power of communication understood as a fundamental human process that links together a speaker and a receiver seeking for common decisions, consensus, and reciprocity (Pasquali 2003).[5] Bertold Brecht attributed this communicative potential to the radio, Jean d'Arcy (1977) did the same with satellite communications, and, more recently, similar considerations have been elaborated about television (Norris 2001).

The Internet is no exception to this scenario. From the beginning of its evolution, the Net has been hostage of the paradox of being born under military authority while representing, at the same time, an unprecedented tool of connection enhancing freedom of association and expression. This paradox led to the polarization of comments on the Internet democratic potential around two opposite perspectives: *cyberpessimists* against *cyberoptimists* (Norris 2001) and *utopians* against *dystopians* (Katz and Rice 2002), with some possibility of finding *cyberskeptics*. For cyberpessimists the Net is nothing but a tool to exacerbate existing inequalities; for cyberoptimists it provides a solution to cross-sectoral imbalances; for skeptics nothing will change. In this context, several authors have called for a more holistic approach that focuses on the potential of the Net without underestimating the risks and inequalities that are present in the virtual space (Katz and Rice 2002; Mossberger, Tolbert, and Stansbury 2003; Norris 2001).

Certainly, communication on the Internet presents several "democratic characteristics" (Simon, Corrales, and Wolfensberger 2002). In the first place, it allows fast, low-cost, and boundless communication that fosters the creation of groups around common interests and ideas. Thus, the Internet empowers individuals and small institutions in a variety of different ways, and, in this sense, it challenges traditional balances of international relations. Furthermore, relationships built upon the Net can be characterized by high levels of intimacy and consolidate membership feelings (Cerulo 1997; Cerulo and Ruane 1998). When the unprecedented, potentially unlimited, and relatively unbounded communication possibilities provided by the Internet met the crisis of traditional politics, rhetoric on the Internet providing a new, alternative public sphere that spread as an easy consolation to the loss of political reference points. However, as soon as the World Wide Web and personal computers entered our daily lives, it became clear that inequalities and limitations already existing offline were affecting also the virtual space, which was far from being a universal and representative space (Poster 1995). Empirical investigations showed that, as it happened for the radio, the satellite, and the television, the sole provision of the Internet technology did not translate automatically into democratic communication or into politically engaged behaviors (Papacharissi 2002; Polat 2005). As Calhoun puts it, "which of the possibilities opened by the Internet are in fact realized will depend on human choice, social organization, and the distribution of resources" (1998:382–383).

Recent discussions about the actual possibilities provided by the Internet to enlarge the public sphere further cooled down the most optimistic views. In spite of the large vitality of the virtual sphere and the numerous *constellations of groups* (Wellman et al. 1996), *communities of practices* (Sassen 2004), *digital formations* (Latham and Sassen 2005), and *enclaves* (Calhoun 1998), several critical factors have been pointed out. The unprecedented amount of information provided through the Net does not necessarily become knowledge (the real key resource) because people have limited processing capacities and because the quantitative increase in information availability does not necessarily guarantee quality (Papacharissi 2002; Polat 2005). Second, communication is such a multifaceted phenomenon that it is impossible to think that the Internet can impact all communication modes in the same way (Diani 2000; Polat 2005). Furthermore, there is an overall agreement on the fact that computer-mediated communication can hardly *substitute* for face-to-face relations, and, in the best case, it ends up supporting them, not substituting for them (Calhoun 1998; Diani 2000; Papacharissi 2002; Polat 2005; Van Dijk 1999; Boase et al. 2006). In general, then, the Internet "democratic" potential seems limited "because of its unequal distribution, highly fragmented structure and increasing commercialization" (Polat 2005:450; see also Di Maggio and Hargittai 2001). The Internet might provide, then, a public *space* but does not constitute a public *sphere* (Papacharissi 2002:11).

There is one major element that emerges from this overview. Most of the reflections developed on the role of the Internet are based on analyses that start from *comparison* between the online and the offline spaces and not on an autonomous investigation of the two. Exploration of online dynamics is mainly driven by the search of how offline mechanisms reproduce or modify while the specificities of the virtual space are overlooked. In the best cases, specificities of virtual interactions are analyzed in terms of the advantages or losses they bring to offline processes (see, for example, in the field of social movement studies how Internet communication benefits the organization of street manifestations in Ayres 1999; Diani 2000; Postmes and Brunsting 2002). Much less attention has been paid to the identification of those logics and strategies that develop online and structure it as a political space (for an exception, see, for example, Caiani and Wagermann 2009). This is certainly due to the fact that the offline dimension, especially when it comes to politics, is still dominant. But this does not imply that the online dimension cannot be understood as a space itself within which conversations are carried on through specific modalities and schemes, with specific aims and through specific logics that can and should be investigated systematically. The extent to which online dynamics reproduce, innovate, or bypass offline processes remains an empirical question that can be answered only if a sound examination of the online space is pursued (Rogers 2010).

Also, it must be said that acknowledgments of online complexity are seldom accompanied by empirical efforts to explore the different forms of online communication. Analysts are aware that the Internet is not a monolithic space: different modes of communication are entailed in different devices. For example, Web communications respond more to the publicity and accountability of online subjects; personal e-mails tend to foster fast communication between limited numbers of people; mailing lists create groups around core issues; blogs have come to represent a mixture of personal, emotional communication, on the one hand, and publicity on the other. However, analyses have concentrated on one device at the time, be it hyperlink strategies on the Web (Della Porta et al. 2006), mailing lists (Calderaro 2008), or newsgroups (Papacharissi 2004). In all these cases, while acknowledging that the device chosen is neither the most important nor the sole device serving a certain function, comparative studies have seldom been realized. In other words, the complexity of the virtual space is acknowledged but the challenges implied by this condition have hardly been faced (for an exception, see Rogers 2010).

3.3 A Framework for the Analysis of the Internet Governance Domain

Pulling all the strings together, our framework of analysis for the study of dynamics deploying in the IG domain thanks to the IGF process results from crossing the two dimensions just illustrated (see Table 2.2). Consistently with the twofold relevance of discourses for the IG domain (reduction of thematic uncertainty and reduction of procedural uncertainty), the first dimension in the table looks at what general types of relations are established between actors in the field: on the one side, semantic relations to look at how conceptual boundaries of the IG domain are progressively shaped (i.e., agenda definition); on the other, social relations to look at how cohabitation patterns are established by actors populating the domain (i.e., political collaboration). Both agenda definition and political collaboration are investigated both offline and online in order to reach a more accurate understanding of the multiplicity of dynamics that are characterizing the redefinition of the IG domain. Within each of these spaces, we chose to investigate those relations that were,

Table 2.2. Framework of analysis for studying Global Communication Governance dynamics

		General Type of Relation Established	
		Semantic	*Social*
Spaces for Network Development	*Online*	Online thematic networks	Mailing list networks
	Offline	Offline semantic networks	Offline collaboration networks

to us, more indicative of multiactor dynamics for the collective construction of the IG discourse.

Therefore, our framework provides a guide for a systematic exploration of different networking activities that are being set up in different interaction territories (online and offline) in order to define IG contents and processes:

- *Online thematic networks* depict how the IG discourse develops online between nodes of different kinds (websites, online documents, blogs, etc.) through online links. In this sense, online thematic networks trace the online thematic boundaries of an issue, gathering together the multiplicity of resources that contribute to the formation of the online discourse on Internet governance.
- *Mailing list networks (ML networks)* represent interpersonal conversational dynamics deployed online between those individuals who have subscribed to specific online thematic spaces. The analysis of ML networks allows us to identify conversational dynamics occurring among committed individuals when physical presence is not possible.
- *Offline semantic networks* depict conversational patterns along which different themes are brought together within agendas. Nodes within these networks represent concepts that individuals from governments, the private sector, the technical community, and public interest groups associate with the idea of IG. Hence, offline semantic networks represent the thematic structure that results from the joined contribution to the construction of the IG discourse.[6]
- *Offline collaboration networks* gather social actors operating in the IG domain. Nodes are all of the same nature, and ties exist among them if one particular kind of relationship exists (cooperation, opposition, sponsoring, etc.). These networks can be read in terms of how political cohabitation patterns are being developed between actors of different kinds within the IG domain, hence, in terms of contributions to the very definition of an IG discourse.

The study of all four types of networks provides a complete understanding of how thematic boundaries and procedures in the IG domain are being redefined in the context of interactions that are pushed forward by the IGF. For each of the four governance network types that we derive from crossing the dimensions of the type of relations established and the space where networks are developed, we isolate and map particularly meaningful relations that could give us some insights on how multiactor dynamics pushed forward by the realization of the IGF are contributing to define the IG domain boundaries and practices. We have said above that we start from an alternative entry point to the study of IG. Indeed, in our exploration we do not look at traditional policy-making processes in the domain or at the actual regulation of specific topics within it (e.g., Internet critical resources management, privacy, etc.). Rather, we investigate the IG domain starting from an alternative entry point: the progressive construction of an *IG discourse* that is fostered by the realization of the IGF, that is, a privileged space where the thematic boundaries of the IG domain are being defined together with interactional patterns between actors that have a stake within it. For this specific reason, we chose and operationalized governance networks as our analytic tool for investigation. However, it should not be forgotten that there are other processes that we could analyze in this same domain—for example, more traditional policy-making activities or international treaties definition. We believe that the structure of our framework, which seeks to analyze the social and semantic activity on IG looking both at the online and offline spaces, should be consistently operationalized in order to look also at these other dynamics. There is a relevant semantic and social component also in policy networks, government networks, and collective action networks.

More than this, we also believe that this framework, the classification of forms of communication networks it builds upon, and the background theoretical approach based on frames and connections can generate positive advancement in the study of GCG dynamics in many other domains. One caveat must be made, though: in switching to other communication network forms and to other domains, the meaning of the social and semantic network traced online and offline will change, together with the relation we will choose to map and very much consistently with the collective purpose that will be produced in the end. To fully exploit the potential of frames and connections through networks, it is then necessary to avoid generalist uses and, as we suggested before, characterize immediately the type of interactions we are examining.

In this sense, our application of the framework in Table 2.2 is just one of the possible applications we could make, and the story of the collective construction of a new, composite IG discourse is one of the many GCG stories that we could tell according to this perspective, thus applying our conceptual starting points. But it is time now to discuss the specific domain we have chosen to explore in more detail, that of Internet governance.

NOTES

1. It must be said, though, that in the attempt to analyze existing differences between systems of political interdependency, there have also been attempts to classify diverse types of policy networks. For example, in his work, Rhodes elaborated a typology that gathers on a continuum five different types of policy networks: policy community/territorial community; professional networks; intergovernmental networks; producer networks; and issue networks (see Rhodes 1996 and, for a summary of the typology, Rhodes 1997). These five forms of policy networks follow one another according to the level of integration between their parts: in this sense, policy communities stand on the one side of the continuum as the most integrated forms of coordination while issue networks stand on the other as the loosest. Yet, the author himself acknowledges that the logic followed to organize the sequence of types is rather relative, with the sole exception of the two opposites (see Rhodes 1997:10).

2. The network concept we lean on here is based on the idea that one set or more of nodes are tied by one type of relation at the time. However, we acknowledge that there is a growing interest in multimodal, multiplex networks, that is, for multidimensional networks where different sets of nodes are joined by different types of relations at the same time (see Contractor, Monge, and Leonardi 2011).

3. It should be specified that actors' nature is conceived here in a general way and very much relies on the institutional perspective provided by the Panel of Eminent Persons on United Nations–Civil Society, according to which there are four main groups of actors participating in the United Nations (UN) system: (a) state and governmental actors, which also include parliaments, associations of parliamentarians, and local authorities; (b) the private business sector, which includes private sector firms, business federations, foundations, and private media corporations and federations representing them; (c) the civil society sector, including mass organizations representing a specific population sector, trade-related organizations (such as trade unions and professionals associations), faith-based organizations, academia, public benefit NGOs, social movements, and campaign networks; and (d) the global public opinion, although this can be conceived as part of the civil society sector (United Nations High-Level Panel on Civil Society 2004). Homogeneous sets of actors are those formed by different entities coming from one of these sectors, while heterogeneous sets involve members of at least two sectors. The underlying (and admittedly rough) assumption is that being a governmental representative, a private actor, or a public interest entity already makes enough difference when it comes to the participation in multiactor processes, both in terms of perception and in terms of possibilities of action.

4. The systematization provided in Table 2.1 should not be interpreted as *opposing* the four concepts but, rather, as analytically distinguishing them, thus attributing

to them a more specific and contextual meaning based on the different way of putting actors into communication relations.

5. Communication is about dialogue and links to the ontological category of *community* (reciprocal action between agents). In this sense, it differs from information, which is rather ontologically connected to the *causality* category (relation between cause and effect) and to the attempts of a speaker to obtain a certain effect, immediate or remote, on a receiver (Pasquali 2003).

6. In this sense, offline semantic networks recall the idea of "concept network" proposed by Carley (1997) and by Diesner and Carley (2005).

Chapter 3

Introducing the Internet Governance Case

It is now time to narrow our focus from the broader field of governance mechanisms of information and communication issues to the Internet governance (IG) domain and to investigate the details of contents and processes that are being developed thanks to the realization of the Internet Governance Forum (IGF). The journey so far has given us the chance to become more familiar with the overall Global Communication Governance (GCG) field and to introduce the theoretical tools we can use to improve our research activities (i.e., frames and networks). In this sense, the first two chapters of the book have worked both as an introduction to the set of problems we will face in looking at the IG case and as a preparation for empirical analysis.

In Chapter 1 we outlined the context for the transformations in the GCG field, where IG emerged as a distinct domain for discussion. We mentioned that IG matters have recently become one of the most important terrains of confrontation where institutional experiments oriented toward multiactor collaboration, such as the IGF, are being played out. We also pointed out that the IGF originated as one of the outputs of the United Nations (UN) World Summit on the Information Society (WSIS) process. After reviewing the structural and historical developments of the GCG field, we concluded Chapter 1 by outlining some research questions that are guiding our exploration of the current phase of debate on the IG domain along the content and process dimensions. From the point of view of contents, we wondered if the emergence of the IG issue in the GCG field carried with it a restriction of the political agenda to information and communication issues and also an enhancement of technicality in the overall political discourse. From the point of view of processes, we wondered if the realization of the IGF could lead to a possible redefinition of roles for governmental and non-governmental actors thanks to the establishment of an open political opportunity structure (i.e., the IGF) for confronting opinions and visions. Who is taking advantage of these opportunities? In which way? With what consequences?

At the end of the first chapter we also pointed out the necessity of developing an ad hoc approach to answering all these questions. In Chapter 2 we introduced the theoretical premises and illustrated the analytical framework we adopt in this book to study the IG case. We reviewed the two main theoretical elements that ground our analysis (frames and networks) with specific attention to their application to the study of political dynamics. Thus, we argued that, instead of looking separately at them, we would benefit from their joint analysis. We therefore adopted an extended conceptualization of *discourse* (see Donati 1992) to summarize both the progressive development of the IG agenda and the consolidation of political relations between institutional and non-institutional actors in this domain. We proposed to look at the collective formation of a new IG discourse in two analytically distinguished, but actually intertwined, spaces: the offline and the online. We therefore identified four types of networks that can be traced and analyzed to gain useful insights on the consolidation of an IG multiactor discourse: *online thematic networks*, which depict how the IG discourse develops online between nodes of different kinds (websites, online documents, blogs, etc.) through selective link strategies; *mailing list networks*, which represent the interpersonal conversational dynamics deployed online between those individuals who subscribe to specific online thematic spaces; *offline semantic networks,* which depict conversational patterns along which different themes are brought together within the IG agenda; and *offline collaboration networks,* which gather together social actors (whether individuals or organizations) collaborating in the IG domain.

In this and the following chapters we apply the analytic framework we outlined in Chapter 2 to the study of IG discourse formation. However, so far we have not described in detail the overall context of the IG domain, where our study is located. We have linked IG and the IGF with previous information and communication debates, but we have not looked at the evolution of the domain itself. We know that the IG issue exploded in the WSIS context like a bomb, but we do not know yet *why* this happened, what were the main positions endorsed by WSIS participants, what were the main claims and the contested points, what attempts were made (before the IGF) to break deadlocks, or what were the principal characteristics of the IGF as a space for multiactor dynamics.

This chapter tries to answer all these domain-related questions. We begin by reviewing the history of IG before and after the WSIS. We then move to clarify the general contours of the current institutional environment where open discussion is taking place, that is, the IGF. While looking at the IGF context, we will pay specific attention to the small-scale multistakeholder experiments that stemmed from the larger process, namely, the Dynamic Coalitions (DCs), which are informal groups that gather voluntary actors from governments, the private sector, and civil society organizations and that are aimed at shaping common discourses on specific issues within the overall IG framework (e.g., freedom of expression, spam, privacy and security, etc.). Our interest in DCs is both substantial, because of the role they played in fostering multiactor collaboration, and functional, as they provided the starting point for our mapping of IG dynamics in this work. In concluding the chapter, some methodological considerations will be articulated to provide an informed premise to the illustration of results obtained from the application of our analytic framework to our case study.

1. ONCE UPON A TIME AND ONCE UPON A POLITICAL PROCESS: INTERNET GOVERNANCE IN HISTORICAL PERSPECTIVE

As we did in Chapter 1 in relation to the broader GCG field, we can tell the IG story from the two intertwined perspectives of content and process.

Looking at the former, we easily see that the story of IG follows the expansion of the Internet worldwide and becomes richer as Internet uses multiply and customize. When the Internet was just born and it involved only a few nodes, IG discourse was mainly revolving around technical matters (such as the development of technical standards and the management of those core resources, e.g., Internet Protocol [IP] addresses, the Domain Name System [DNS], and the Root Server System) that were fundamental to the system's functioning. At that stage, technical standards organizations and laboratories were the privileged sites for carrying on a conversation that, overall, was quite exclusive in terms of the themes and actors involved. But the more the Internet system grew, stimulated by and stimulating new uses and activities, the less the discussion on *how* it could be managed and even on *what* had actually to be managed (i.e., only technical aspects or also uses and misuses of the Net) could remain limited to technical issues and locations.

Hence, thematic changes generated modifications also at the level of processes. Indeed, if at its very beginning IG was translated into practice through self-governance practices carried on by technical bodies, when the Internet became a worldwide phenomenon, then other actors and stakeholders became (or wanted to become) involved. Besides technicians, governments, and intergovernmental organizations (IGOs), a growing number of non-governmental entities (both from the private sector and representing public interests) entered into the discussion. The increased number of stakeholders willing to take part in the governance discussion soon revealed both the impossibility of carrying on a discussion based merely on technical matters and the insufficiency of technical forums for hosting an overall and overarching debate.

Hofmann proposes to consider IG as a "process of searching that has unfolded in several stages" (2006:2). She identifies three phases character-ized by different "spheres of activity, constellations of actors, policy agendas and perceived problems" (Hofmann 2006:2): the *technical regime*, going from the early days of IG to the mid-1990s; the *institutionalization of self-governance phase*, from the mid-1990s to the beginning of the WSIS in 2003; and, finally, a "forum shifting" phase, which is currently ongoing and during which actors and existing governance settings are being redefined. Although other historical reconstructions (e.g., Kleinwächter 2004, 2007) do not single out neat phases of development, they highlight the same evolutionary pattern, which includes an initial technical stage; subsequent attempts to progressively institutionalize self-governance experiments; and, finally, a shift during the WSIS after which IG is consolidated on the global stage and the discussion on possible reform patterns in this domain is speeded up.

In the next sections we summarize the main developments of the IG dis-cussion before the realization of the IGF.[1] The first part reviews how IG developed before the realization of the WSIS; in this sense, it joins together the two phases that Hofmann (2006) calls the *technical regime* and the *self-governance* phases. Secondly, we will retrace the main developments of the IG discussion within the institutional space provided by the UN, up to the beginning of the IGF in 2006.

1.1 Technical Regime and Self-Governance in the Internet Governance Domain

Unlike what happened in all other telecommunication domains such as tele-phone, the early days of IG saw the overall absence of national governmental and intergovernmental regulation and interests (Kleinwächter 2010). Many of the bodies that today are still regulating Internet development were estab-lished in this first phase, which Hofmann (2006) calls the *technical regime phase*. However, IG institutions did not involve governmental actors, who, in fact, tended to overlook this domain (with the exception of the United States, which had financed the Advanced Research Projects Agency Network [ARPANET] project from which the Internet we know developed).

In this initial stage, the Internet system was working mainly under the guidance of the Internet Engineering Task Force (IETF).[2] The IETF is "the protocol engineering and development arm of the Internet"[3] and was officially established in 1986 (after it existed informally for a while) in connection with the Internet Architecture Board (IAB).[4] Within the IAB, an open community of designers and technicians, whose membership in the institution is determined on the basis of their expertise, is appointed to develop Internet standards. Furthermore, while the extension of the Net was still limited, the various management tasks of establishing the core resources for communication flows were carried on efficiently and pretty much informally through the Internet Assigned Number Authority[5] (IANA) and the work of a single man, Jon Postel, who was coordinating the assignment of IP addresses and of Internet names. However, when the Internet began to grow rapidly in the 1990s thanks to the development of the World Wide Web (WWW), its governance system had to develop flexibly and rapidly, yet without changing the overall bottom-up approach to management that was adopted in the beginning. In 1992, developers, activists, and experts founded the Internet Society (ISOC),[6] a nonprofit "professional membership organization of Internet experts that comments on policies and practices and oversees a number of other boards and task forces dealing with network policy issues."[7] The establishment of the ISOC was meant to be an explicit signal that, even when IG matters began to exceed purely technical issues, the search for solutions was still carried on within the Internet community, thus avoiding conventional top-down decision-making mechanisms.

The absence of governmental actors in the overall IG picture was not perceived as a lack but rather as one of the reasons for Internet success, as it avoided "time-consuming and costly procedures which would reduce the speed of innovation on the Internet and block the creation of new Internet services and applications" (Kleinwächter 2007:42). IG was in fact seen as a successful story of "self-governance," in contrast to the intergovernmental regime that was characterizing other communication and information domains (Hofmann 2006:9) and, also, when considering the increasing difficulties that institutional actors were starting to experience in those days (see Chapter 1). Despite this enthusiasm and the apparent effectiveness of self-regulation practices, a purely technical development and management perspective could hardly hold up against the spread of the Internet and the consequent multiplication of issues that became connected to it. Indeed, with the development of the personal computer and the advent of the WWW, the Internet exited from military and academic laboratories and entered houses and firms, innervating reorganizational processes and providing the backbone for the development of global communications. Economic revenues deriving from the exploitation of the Net became evident pretty soon, and governments started to realize that they could not ignore the Internet any longer. In this context, the self-governance arrangements that had been consolidated since the birth of the Internet could hardly cope with the pace of the expansion of online activities. With the Internet as a mass medium, the coordination and allotment of core resources—in particular of domain names and IP addresses[8]—became increasingly complicated. Furthermore, the technical core of the IG discourse began to be enriched by other difficult, controversial, and socially connected topics, such as the protection of personal data or the digital divide (Hofmann 2006). If these thematic developments did not lead immediately to a change in the bottom-up governance approach, they nonetheless helped to put existing mechanisms under scrutiny and opened the floor to a discussion of their renewal, yet without shifting toward a more formal top-down governmental regulation (Hofmann 2006).

However, tensions arose in relation to technical matters, especially matters pertaining to the expansion of the domain name space. Not only was this operation considered necessary to meet the increased usage of the Internet, but domain management also appeared as a very remunerative business (Kleinwächter 2004). Network Solution Inc., a private company that was managing .org, .net, and .com top-level domains (TLDs), strongly opposed any attempts to enlarge the system without its direct control. Jon Postel and the technical community had elaborated in the meantime an alternative solution that entailed a more marked involvement of governments and intergovernmental institutions. In 1996, the International Ad Hoc Committee (IAHC) was created to deal with the problem of TLDs. As a result, the Internet technical community (represented by ISOC, IANA, and IAB), intergovernmental organizations (represented by the International Telecommunication Union [ITU] and the World Intellectual Property Organization [WIPO]), and the business community (represented by the International Trademark Association [ITA]) converged on the draft of an agreement that aimed at expanding the TLDs in a coordinated way. In 1997, negotiations concluded with IAHC members signing a "Memorandum of Understanding on General Top Level Domains." However, this document encountered the opposition of the U.S. government, which, in turn, started to walk the IG road alone and drafted a Green Paper. This alternative document generated reactions from other institutional actors and from the Internet community, who both feared and opposed the overwhelming role played by the United States in the Internet system. Despite general mistrust and opposition, in 1998 the Green Paper turned into a White Paper and set out the principles for an effective IG (stability, competition, private bottom-up coordination, and representation), thus foreseeing the replacement of the traditional technical supervision with control by a formal nonprofit organization based in the United States. It was the birth of the Internet Corporation for Assigned Names and Numbers (ICANN).

ICANN was not created to perform technical or standardization functions but rather to represent all actors in the Net, from the technical community to the users passing through private and institutional interests, when it comes to the allocation and management of Internet resources. The original organizational plan for the ICANN maintained the traditional bottom-up approach to governance through the establishment of a decision-making Board of Directors formed by private sector representatives, the technical community, and civil society. Governments were also included in the picture within the Governmental Advisory Committee (GAC), whose role was meant to be only advisory, as suggested by the name. Since its early days, ICANN faced many difficulties, and many objections were raised, both against its structure (particularly in relation to the election of users' representatives in the board) and against its functioning mechanisms (particularly with reference to the management of the DNS and of Internet infrastructures) and its financing strategies (Hofmann 2006; Mueller 2002). In general, the main issue with the ICANN was the difficult task it had been assigned, namely, joining together "the need for technical coordination to regulation of the industry built around the resources it manages" (Mueller 2002:218). Observers pointed out that such a task was too broad and too complicated to be pursued by one single organization, especially since its multiactor composition did not provide any actual guarantee for transparency or for equal representation of all stakes in the play. Despite a formal inclusion model, which brought all stakeholders into the same organizational frame, the representation of interests was substantially unbalanced: economic interests were considered as predominant in the work of the organization, even though organizational settings did not allow firms to fully exploit their advantaged market position. Conversely, users were progressively left behind in comparison to firms but also to governments, who, in fact, saw their role strengthened in the aftermath of the Twin Towers attack in 2001 (Kleinwächter 2004). At the dawn of its third year of existence, ICANN tried to apply organizational constraints to these multifaceted problems and pushed for a more solid private-public partnership between firms and governments, thus abandoning the pure self-governance model and shifting the main IG focus from cyberdemocracy to cybersecurity (Kleinwächter 2007:54).

1.2 IG@UN

The shortcomings of self-governance methods led to a shift in the way IG was conceived and played out. In a context where a spreading sense of mistrust was affecting the sustainability of self-governance experiences, governments and intergovernmental organizations were not considered any longer as obstacles but rather as necessary complements of the more traditional bottom-up governance approach. Their involvement began to be perceived as necessary, especially to address the social and political implications of technical matters, which could no longer be avoided. However, governments refused to take the lead within IG organizations, and, while acknowledging their commitment to play an enhanced role in the domain, when the WSIS began they managed to bring IG into a "new, intergovernmental territory" (Hofmann 2006:14). Still, this shift in location did not immediately renew the debate. Quite the opposite: during the preparatory process to the Geneva meeting, IG was hardly a popular topic, and, with only few exceptions, it was almost ignored in all documents (Kleinwächter 2004:44–49). Furthermore, on the side of IG institutions, ICANN and other IG organizations were paying very little attention to the development of the WSIS (Kleinwächter 2004).

Such an initial and mutual inattention between the traditional and the intergovernmental environments of discussion did not last long. As we have seen in Chapter 1, the first phase of the WSIS aimed at shaping a shared vision of the information society and at identifying directions for its practical realization. In this overall context, the Internet was occupying a paramount role and, therefore, its management (*how* and *what* to manage) became a crucial thematic cornerstone in the whole discussion. Although the official document draft did not initially contain any systematic reference to the management of the Net, the open and participatory features of the Summit provided a multiplicity of actors with the opportunity to speak their minds on these issues. When some governments (especially China and other so-called "developing countries") explicitly opposed the dominant role of the ICANN and the U.S. government and proposed to bring IG matters under the inter-governmental umbrella of the ITU (Kummer 2004; Peak 2004), the conflict over IG exploded like a time bomb. This proposition encountered the strong opposition of the United States and the European Union (EU), which instead insisted on the necessity of maintaining the status quo built on the ICANN's central role. The fact that different stakeholders had all different opinions on what IG meant and, therefore, made different proposals to renew its mechanisms and guidance worsened the harshness of the leadership issue. The diversity of position was a feature separating not only the three sectors (institutions, private sector entities, and civil society) but also the actors grouped within the different constituencies (Peak 2004). Those who were defending the status quo tended to adopt a more technical perspective for discussion, whereas actors calling for a substantial reform of governance mechanisms pushed forward a wider vision of IG that included public policy and social issues. Consistent with the overall organizational trend adopted by the civil society sector to channel competencies and thematically focused inputs, an

Internet Governance Caucus (IGC) was established. The IGC gathered various representatives from non-governmental public interest entities that had a marked expertise in relation to the issue and that, since its establishment, had provided important contributions on different aspects, such as policy matters connected to the governance of the Internet, the necessity to enhance multistakeholderism in the domain, and all the various problems connected with ICANN functioning mechanisms. The IGC soon became a reference point in the articulation of an alternative road to governing the Net (Peak 2004), while, overall, the discussion on IG rapidly became inflamed. By the end of the first phase of the Summit, the disagreement on the IG topic was general, crosscutting membership sectors and proving difficult to resolve quickly. The only plausible solution was then to "agree to disagree and postpone" (Kleinwächter 2004:48–50).

A further occasion for reflecting on possible improvements on the IG terrain was provided with the institution of a UN Working Group on Internet Governance (WGIG).[9] The bases for the work of this multistakeholder group[10] were posed at the end of the first phase of the WSIS (WSIS1) by the WSIS Declaration of Principles and the Plan of Action, according to which the WGIG had the task of developing a working definition of IG; identifying public policy issues converging in the IG domain; developing a common understanding of the roles and responsibilities held by different actors (governmental and non-governmental, from both developed and developing countries); and preparing a report on the results achieved that would work as the basis for the second phase of the WSIS (WSIS2) (WSIS 2003:para. 13). This composite task was far from easy, but the group managed to accomplish it. As far as the elaboration of a working definition, the WGIG agreed to conceive of IG as "the development and application by Governments, the private sector and civil society, in their respective roles, of shared principles, norms, rules, decision-making procedures, and programmes that shape the evolution and use of the Internet."[11] The abstract nature of this definition provided a common ground for further negotiations but was compensated by the identification of a list of concrete policy issues necessarily linked to IG.[12] As far as the roles and responsibilities of actors are concerned, the WGIG set the direction for multiactor cooperation in this domain: governments were assigned mainly policy-making and oversight functions; the private sector was recognized to be in charge of developing best practices, fostering innovation, and building capacity; and, finally, civil society was assigned the role of bringing into IG processes more expertise, defending the public interest, and fostering the diffusion of innovation and best practices (WGIG 2005:para. 29–34).

In analyzing the value of the WGIG experience, Drake (2005) distinguishes between *procedural* and *substantive* contributions. From the point of view of the former, the WGIG demonstrated the benefits of multistakeholder collaboration, actually facilitated WSIS negotiations in the Tunis phase, provided a common ground for joint work, and promoted public engagement in the IG debate. In this sense, it has been underlined that the multistakeholder and flexible structure of the group actually "produced a new culture of interaction" (Kleinwächter 2010:81): in lowering the ideological level of the discussion while searching for concrete solutions to be adopted through consensus, the WGIG actually showed that multistakeholderism can be translated into effective practices, thus grounding collective decision-making processes (not only serving brainstorming or agenda-setting functions). As far as the substantive contributions are concerned, it must be said that the WGIG managed to accomplish all the tasks it was assigned by the WSIS documents. In the first place, it demystified the nature and scope of IG through a more clear-cut definition of the meaning of *governance* ("[it] does not refer to government but rather to the act of steering"; see Drake 2005:255), and it provided a broad working definition that served as a base for developing further collaborations. Moreover, its work was fundamental in identifying those technical and socio-political areas that need to be taken into consideration when dealing with IG and when outlining several alternative models to manage critical Internet resources.[13] Finally, the WGIG bears the responsibility of having proposed for the first time the establishment of the IGF as a new, open, and participatory space to stimulate discussion and understanding of IG (Drake 2005:254–265).

Certainly, while the WGIG experience helped diminish fears of a UN takeover on IG, its results also provided the necessary starting points for negotiation during the WSIS second phase (WSIS2) in Tunis. Starting from these bases, in the Tunis Agenda (the final document of the WSIS Tunis phase), IG was recognized as being a complex domain requiring further multiactor collaboration. With the end of the WSIS, the process leading to the realization of the first IGF meeting officially took off.

2. THE INTERNET GOVERNANCE FORUM

The IGF stands in the IG landscape as a peculiar space where all interested actors, whatever their nature, can enter and contribute on an equal footing to a collective dialogue on one of the most relevant topics within the GCG field. Its structure and functioning mechanisms were designed to constitute an "embryo of a *Governance for the Internet Age*" (de la Chapelle 2007:25, emphasis added). The IGF constitutes only one part of the IG system and is not meant to be the place where actual decisions on how to reform it are made. Rather, it provides an "incubator of ideas" and a "platform for dialogue" (Kummer 2010:iii). Indeed, the IGF is not a decision-making body but, rather, a process that builds on dialogue. However, even though it has no traditional policy-making task (not even in the "lighter" version that was experimented with at the WSIS with the drafting of the shared vision and the plan of action), the IGF has a twofold hard task to accomplish: defining the boundaries of the IG domain and translating the principle of multistakeholderism into effective practices to ground the reform of global governance arrangements within this complex thematic area. According to the Tunis Agenda, where the IGF mandate and main features are outlined, the IGF constitutes a space, first, to foster a dialogue among stakeholders on old and new issues relating to IG and, second, to articulate along different lines the IG concept. More generally, the IGF aims at filling the vacuum with a specific space in which "homeless" issues, which do not fall under the competencies and the legitimate range of action of existing bodies, can be debated. In doing so, it promotes the engagement of various different actors in the different aspects of IG, thus stimulating a participatory debate on old and new infrastructures, competencies, resources, and competencies divides (WSIS 2005:art. 72).

It has been argued that the IGF was not initiated because of the urgency to find an innovative way to deal with IG matters but, rather, as a compromise to avoid a lack of consensus at the end of the whole WSIS process (de la Chapelle 2010). Also, some observers argue that the very proposal for the realization of a forum specifically focused on IG issues originated as a compromise already within the WGIG because, however good multistakeholder collaboration might have been in that case, it was impossible to reach an overall agreement on how different groups had to be involved in global policy development and decision-making procedures (Kleinwächter 2010). Compromise or not, the IGF currently represents an unprecedented and peculiar institutional effort to focus specific attention on a set of controversial issues that give rise to many symbolic conflicts between old and new political actors, not only because of their inherent relevance but also because they touch on a plurality of different and opposite interests.

The better way to realize the potential of this process for the renewal of the GCG field is to examine its specific and unique features, which help to shape it as an open political opportunity structure for all stakeholders. The first peculiar trait of the IGF is its format. Indeed, the IGF is shaped as a process and not as an annual event (de la Chapelle 2010); it lasts for 5 years, and its mandate can be renewed by the UN General Assembly. The whole process is punctuated by annual meetings, but it also flows in between these points, thanks to the annual open consultations and meetings of the Multistakeholder

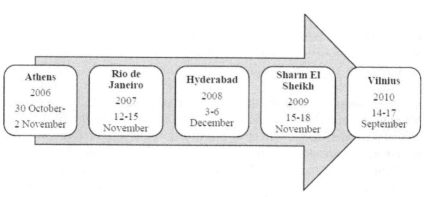

Figure 3.1.

Advisory Group (MAG), which is a sort of "program bureau" for the whole forum process composed by governmental, private sector, civil society, and technical community actors. Officially convened in 2006 for the first time, the IGF closed its first round in 2010 and, over the years, has followed the meeting list illustrated in Figure 3.1.

The idea of having a 5-year process is justified in the first place by the task assigned to the Forum: fostering a genuine multiactor dialogue on IG themes and actors' roles is easier said than done. Indeed, it implies first of all a clarification of boundaries. During the WSIS process it was evident that there was no agreement on *what* should fall under the IG label. Moreover, the definition provided by the WGIG represented a common denominator that served to further negotiations and, in the end, did not reduce the overall thematic uncertainty characterizing the IG domain. Nor did the WGIG definition explore in depth the level of procedural uncertainty, as it provided only general benchmarks that needed to be operationalized more precisely.

To perform its twofold task, the IGF was relieved of any decision-making function so as to allow for a free and possibly unconditioned peer dialogue. The lack of commitment to the production of binding results constitutes then the second peculiar trait of this institutional process. When the idea of the IGF started to be formulated after the first phase of the WSIS, its proponents were well aware of the fact that there are multiple places where Internet policies are taken—from IG institutional bodies such as the ICANN to intergovernmental entities such as the EU to national governments. Therefore, the IGF was not conceived as a duplicate but, more broadly, as an institutional space wherein different viewpoints and experiences could be exchanged to prepare the terrain for informed, coordinated, and shared policy-making activities (Kummer 2010). It is in this sense that the Tunis Agenda defines the IGF as a "*neutral, non-duplicative* and *non-binding* process" (WSIS 2005:art. 77, emphasis added): although it might overlap with other IG policy forums in terms of issues discussed, its non-binding and open nature prevents formal interferences and deadlocks, thus avoiding the risks and the burden of negotiations over final official outputs (Kleinwächter 2010; de la Chapelle 2010). In sum, far from launching open challenges to existing governance mechanisms for the Internet, the IGF builds on them "with special emphasis on the complementarity between stakeholders involved in the process—governments, business entities, civil society and intergovernmental organizations" (WSIS 2005:art. 73).

The joint collaboration of all stakeholders is facilitated by the rules of access to the Forum, which constitute the third characterizing feature of the IGF. In fact, participation in the process is voluntary and based on a "radical departure from traditional intergovernmental processes" (de la Chapelle 2010:93), as it is grounded in experience and expertise in the domain. In this sense, while "in other UN forums (including WSIS), participants are continually reminded of their status within the meeting as insiders—governments, delegates or representatives of IGOs—or outsiders—private sector and civil society—[in the IGF] there was no discrimination between participants according to the status of stakeholder groups" (Esterhuysen 2008:38). Observers agree that this unprecedented openness is a merit of the IGF format because it allows for inclusiveness and facilitates dialogue by eliminating those formal constraints to discussion stemming from status-based issues. In conjunction with the non-binding character of the discussions, the Forum openness has increased the chances of having it accepted by all different constituencies, as it is perceived as a possibility for mutual exchange and learning, in the process wiping away fears tied to the formal inclusion of civil society and non-traditional IG actors.

A fourth peculiarity of the IGF is that, to better exploit the potential of open participation, the Forum structure is flexible and can be subjected to changes. The MAG is in charge of defining the general agenda and the specific structure of annual meetings, and, so far, this task has been carried out by building on the experience developed on the ground year after year in a cumulative learning process. In this sense, there are two main modifications that seem to reflect the results generated by repeated interaction between actors within the IGF space. In the first place, from the point of view of process, while at first the main sessions were basically the illustration of insights provided by experts in front of a large assembly, they turned into open dialogues where experts and representatives of groups and the public interacted directly (mainly through a question-and-answer mechanism) with the help of a very basic moderation. This enhanced interaction is the product of an increased propensity to discuss issues directly, which, in turn, derives from an overcoming of initial difficulties to set up mutual communication relationships and reciprocal trust. Secondly, from the point of view of themes, the present triangulation of the main thematic pillars discussed though open dialogues in the IGF meetings (i.e., access and diversity; security, privacy, and openness; Internet critical resources) is a result of a progressive refinement of the overall agenda. In fact, when the IGF started, security, openness, diversity, and access were discussed separately in the so-called "SODA agenda" (de la Chapelle 2010), while they are now clustered together to recognize the intertwined and multilayered quality of the discussions developed on the ground over the years. Moreover, in an attempt to duplicate discussions going on elsewhere, critical resources were not explicitly included in the first meeting in Athens, but they were reintroduced after participants lamented their absence.

And yet, if these procedural and thematic modifications, which were made almost "in real time" and generated directly from the interactional practices tried on the ground, make the IGF a welcome "embryo" of innovative governance, it is not the "perfect" embryo. At the end of the IGF's first mandate, reflections are multiplying about the challenges that remain and that require deeper and more precise tuning at both the process and content levels. A first, major area of discussion concerns the MAG, its composition and its function. Even from the short overview provided in this chapter, we can derive a sense of the importance of the MAG in shaping the thematic and procedural boundaries for multiactor convergence in the IGF space. At present, though, the mechanisms that are regulating its composition and its renewal are hardly perceived as satisfactory: rather, they clash with the overall open and transparent modus operandi of the forum they organize because of the persistence of traditional elements of bureaucracy and closure reminiscent of "old" governance dynamics (de la Chapelle 2010; Kleinwächter 2010; Malcom 2008; Mathiason 2008).

A second critical matter is that of the final results of the IGF process. Although some observers (see de la Chapelle 2007, 2010; Kleinwächter 2007, 2010) argue that the IGF format should be conceived as the main result of the process itself, there is an ongoing discussion about how to measure the outcomes of what has been portrayed as a 5-year learning experience. Over time, the distinction between "forum doves" and "forum hawks" (Malcom 2008), the former pointing to the usefulness of cognitive and ideational outcomes and the latter lamenting the lack of any formal recognition of results achieved, has worsened to the point that some actors define the IGF as a "waste of time" while some others underline the various potentials of the process: as an *observatory* for Internet development; as a *school* where it is possible to learn a lot on the Internet and its governance; as a *laboratory* for testing solutions and proposed arrangements; as a *clearinghouse* for the dialogue between multiple groups of stakeholders; as a *scout* to identify in advance key topics; and as a *space for watchdog* activity (Kleinwächter 2010).

What seems to be missing, in this context, is a systematic analysis of how the open and participatory environment provided by the IGF is leading to the production of meaningful results beyond conventional policy-making activities. Not only could the results of such analysis help clarify the roles that different actors occupy in the development of a collective discourse on IG, but the clarification of roles and functions played by different constituencies could also provide some general guidelines for rethinking institutional arrangements, such as, for example, MAG nomination procedures. More broadly, a systematic analysis could provide some insights for better understanding actual chances and paths to reforming supranational political arrangements toward inclusive and participatory settings. It is true that the IGF has shown that different actors can converge in the same space and that a confluence of ideas, perceptions, and resources can generate results, such as the progressive clarification of IG thematic boundaries. Moreover, the constant refining of the event format to meet exigencies expressed directly by participants is a signal that interactional practices on the ground are leading to an enhanced communication between actors who were not used to talking directly to one another and to a progressive institutionalization (Jepperson 2000) of multiactor politics. Thus, the replication at the regional level of IGF meetings aimed at discussing more localized concerns is further evidence of the success of multistakeholder collaboration (de la Chapelle 2010; Kleinwächter 2010). However, it remains to be empirically assessed what this multistakeholder communication is made of, what kinds of logic are orienting the construction of a common discourse on IG, and who is playing a predominant role within the discursive dynamics in this domain. Furthermore, although discussion and comments are focused on the Internet and observers have put a certain emphasis on the relevance of virtual communication and remote participation in limiting participation hindrances, the dynamics established online have seldom been analyzed to assess how the collective construction of online collaboration contributes to the deployment of a new multiactor IG discourse.

Overall, then, if the impression at the end of the first IGF cycle is positive, it remains a matter of empirical research to assess what multiactor dynamics are generating both in terms of the collective construction of the IG meaning and in terms of the roles played by actors in open and participatory governance arrangements. If observers have welcomed the establishment and the progressive consolidation of an IG network (de la Chapelle 2010) as a metaphor for actors converging in the same institutional space on an equal footing, they end up being too generous if they forget that, albeit horizontal, a network is full of imbalances and, consequently, of power inequalities. The processes that are sustaining the emergence of norms in the IG domain have not been explored in depth yet, and much has been left to say about how these are structured and developing.

A systematic application of the framework proposed in the previous chapter does provide a first meaningful step in this direction, but, before proceeding, it is necessary to spend some more time examining the small, thematically specific groups generated within the IGF context and labeled Dynamic Coalitions (DCs). Our interest in DCs is both substantial, as they allow for practical experimentation with voluntary multiactor collaborations, and functional, as they provide the starting point for the mapping activity pursued in this work. Let's clarify what these groups are and why we chose to lean on them for our mapping efforts.

2.1 Dynamic Coalitions as Thematic and Interaction Proxies

Dynamic Coalitions (DCs) on IG are a further interesting and peculiar element of the IGF. The label *dynamic coalition* was created shortly before the Athens meeting in 2006 to identify informal—albeit recognized—groups of actors of different kinds acting both *within* the IGF space and *in between* IGF meetings to sustain the articulation of a discourse on specific IG themes, such as privacy and freedom of expression. Their name is meant to differentiate them from previous civil society groups established during the WSIS (civil society working groups and caucuses), as DCs are composed not only of civil society entities. Coalitions are, in general, informal groups that reflect the multistakeholder approach initiated within the WSIS process, formalized through the WGIG experience and then articulated throughout the Forum structure. They gather actors from governments, the private sector, and civil society organizations and are aimed at shaping common discourses on specific issues within the overall IG framework. The idea of *coalition* was adopted to suggest the collective effort pursued through the establishment of these groups which was aimed at defining more clearly the conceptual boundaries of IG subdomains and at promoting multiactor collaboration in relation to specific themes that have emerged, over time, as particularly relevant. More particularly, DCs are *dynamic* because they are based on the creation of dynamism between actors of different natures that were not previously unified by any kind of interaction and because they dynamically evolve together with the overall IG thematic boundaries. Thus, these groups are *coalitions*, even though this characterization is quite metaphorical. Indeed, these groups do not correspond properly to conceptualizations of *coalitions* as forms of collectives that join together homogeneous actors in networks with a substantial degree of coordination and exchange over time (Diani and Bison 2004). Rather, they are coalitions in a more metaphorical sense, as they bond together previously disjoined actors under the same thematic umbrella, thus giving to all actors the same status, that of interested party.

Malcom (2008:379–380) classifies DCs into three categories: *networks*, such as the Stop Spam Alliance, which are coalitions serving as coordination groups between already existing programs; *working groups*, such as the Internet Bill of Rights (the former name of the Internet Rights and Principles DC), which allow their members to work together on a joint program; and *BOF* (*Birds of a Feather*), which work as spaces in which people with the same thematic interests can converge yet without an explicit joint program of work. More generally, despite single realizations and peculiarities, DCs can all be conceived as "trans-national information networks" (Singh 2002), as they are publicly recognized multiactor structures operating transnationally and focusing on a variety of issues that pertain to the development, diffusion, and use of communication technologies; they all have a more or less clear program of action, which, if not specifically oriented to the achievement of a task at least tends to provide a space where controversial issues and difficult relations can be constructed; and they all start from the assumption that actors' convergence is thematically driven by a commonality of interests.

When they first appeared on the IGF scene in the immediate aftermath of the Athens meeting, DCs were welcomed as small-scale multistakeholder experiments. From the meeting in Rio on they have also been officially allotted ad hoc time slots within the IGF schedule, with a consequent increase in their activities and membership. One of the reasons why these groups grew up so rapidly is that DCs have no explicitly defined membership criteria or procedural guidelines, the only requirement being their multistakeholder composition. Besides this, any individuals and organizations from all over the world can become DC members as long as they have a clear interest in and competence for the development of a specific thematic area. Most of the interaction between members and between different coalitions is carried on through Internet devices, especially e-mail and electronic platforms, as members are globally dispersed actors that are joined together by common thematic interests. In keeping with their heterogeneous composition, over the first 5 years of the IGF these groups have shown very different levels of activity, and often their multistakeholderism was more formally guaranteed than substantially respected. In fact, in almost all groups, governments and the private sector were less represented than civil society (de la Chapelle 2010).

At the time of this writing, the IGF website reports the existence of 19 DCs that are differentiated between active and nonactive (Table 3.1). Among nonactive DCs, some groups are completely inactive while some feeble mailing list activity still happens in other cases, albeit not sufficiently to revitalize the

Table 3.1. Status and names of Dynamic Coalitions (DCs) established within the Internet Governance Forum process before the Vilnius meeting

Status	Name
Active DCs	1. Dynamic Coalition on Internet and Climate Change
	2. Dynamic Coalition on Accessibility and Disability
	3. Dynamic Coalition on Child Online Safety
	4. Gender and Internet Governance
	5. Freedom of Expression and Freedom of the Media on the Internet
	6. Coalition Dynamique pour la Diversité Linguistique
	7. Dynamic Coalition on Internet Rights and Principles
	8. Dynamic Coalition on Open Standards
	9. Dynamic Coalition on Core Internet Values
	10. Youth Coalition on Internet Governance
	11. Dynamic Coalition on Convergent Media
	12. Dynamic Coalition for a Global Localization Platform
	13. Dynamic Coalition on Social Media and Legal Issues
Nonactive DCs	14. Framework of Principles for the Internet
	15. Online Collaboration Dynamic Coalition
	16. A2K@IGF Dynamic Coalition
	17. Dynamic Coalition on Access and Connectivity for Remote, Rural and Dispersed Communities
	18. Dynamic Coalition on Privacy
	19. The Stop Spam Alliance

Source: www.intgovforum.org.

work of the whole coalition. Thus, there are significant variations in activity rates also among DCs classified as active.

The easiness with which these groups turn on and off does not indicate an overall ineffectiveness. As small-scale multistakeholder experiments, DCs face several limits and constraints that influence their liveliness. In the first place, in being so strictly tied to their thematic focus, they are very much affected by the variation in attention levels given to the thematic area they represent. Although the IGF is an open space, attention remains a scarce resource, and not all matters that appear on the IG agenda are paramount: from time to time priorities change and themes receive more or less attention. The more an issue is appealing, the more it stimulates collective action also through DCs, and, vice versa, the more active DCs are, the more an issue is likely to remain central in the agenda. Secondly, as DCs have no formal mandate, they soon end up depending particularly on the efforts of a few committed individuals. In relying very much on personal and voluntary resources, their activities are easily dissipated when initial energies are not (or cannot be) properly replaced. Moreover, the absence of an ad hoc central mechanism for content and process refinement (like the one that the MAG performs for the whole IGF process) hampers the flexible and systematic adaptation of coalitions to the evolution of the discussion, thus leaving to informal and voluntary initiatives the difficult task of fostering multiactor thematic confrontation. Thirdly, and very much as a consequence of this second point, when voluntary groups like these are not sustained by a neat identification of goals, they tend to cool down quite easily. The lack of a strong shared identity jeopardizes the sustainability of a mobilization effort that is, by nature, transnational and that exploits only to a very limited extent the power of face-to-face interactions to foster reciprocal recognition and group consolidation. For example, one of the most active groups among DCs is the Internet Rights and Principles group. This DC aims at collectively drafting a document to be presented in the institutional gathering of the IGF, and, consequently, it works in the direction of goals, principles, and ideas that can join together IGF participants. Other coalitions, however, do not pursue a concrete task beyond the preparation of annual meetings within the IGF event, and, therefore, the discussion they host is less sustained and lively. Overall, there DCs work as long as they find energies and reasons to do so.

In this sense, the label *coalition* these groups are assigned is not representative of their functions. It is not stated anywhere that they have to sustain the proper formation of coalitional processes between interested parties. Hence, their alternate destinies and varying degrees of activity should not be evaluated by referring only to the proper coalition formation but, rather, in terms of how much they succeed in raising awareness on specific issues or in deepening certain aspects that are missing from the IG agenda. Certainly, some of the inactive DCs ceased their activity because of an overall lack of resources. However, it seems reasonable to hypothesize that activities within specific ad hoc groups tend to cool down if the advocacy task they actually are meant to play is accomplished or, conversely, is impossible to achieve. It is in this sense that the issue-driven nature of DCs reveals its influence: their destiny parallels the popularity of the theme they revolve around. Far from being a failure of the multistakeholderism principle, coalitions just end up being encompassing groups working as smaller "incubators" of ideas and "platforms" where nothing but thematic articulations are produced through communicative exchanges between different kinds of actors. An evaluation of DCs' actions that is based only on the quantity of messages exchanged or on the levels of representation of the three sectors within the group would perpetuate the mistakes connected to the narrower interpretation of multistakeholderism (see Chapter 1). Also in the case of DCs, it is with reference to relations between actors and themes that the work of these groups should be evaluated.

In the context of this research, we do not want to comment on the levels of efficacy of these groups in the IGF framework or more generally, but we consider these groups as *proxies* to map the lines along which the IG domain is consolidating from both a thematic and a procedural perspective. Indeed, DCs share the main features of the broader process they are inserted within: although they have a general thematic focus, their contents horizon is very much uncertain (hence, their thematic clarification task); moreover, they set up open discussion environments that reproduce the overall IGF approach to the promotion of a clarifying and collaborative multiactor dialogue. Looking more specifically at IG contents, then, DCs can be conceived as *thematic proxies* because the whole range of themes on which coalitions focus represents the complexity of the contemporary thematic rainbow associated with the general IG label. From a procedural point of view, instead, they can also be understood as *interaction proxies* because they are transnational networks of IG "activists" who come from different institutional and non-institutional constituencies, mobilize on a voluntary basis starting from their interests and competencies, hold high levels of expertise, and distinguish themselves from "simple" forum participants who do not explicitly commit to carrying on discourses in between official IGF meetings. It is starting from this idea of DCs as proxies that we began to trace the different governance network types contained in our framework of analysis (see Chapter 2) to explore the vast IG domain. Before moving to the illustration of results, though, it is worthwhile to reflect a bit on the methods through which relational data were retrieved, starting from DCs, to build up the different networks and on how these have actually been constructed and prepared for analysis.

3. METHODS FOR DATA COLLECTION AND NETWORK CONSTRUCTION

To map the communication governance networks developed in the IG domain, we must first set the boundaries of the population under examination or, in other words, determine what range of actors and issues will generate the networks that we will trace and analyze. The problem arises precisely because of the open character of the process under examination, wherein any entity and any theme can enter the discourse. How could we keep track of the multitude of individuals, organizations, and their thematic concerns that innervate the whole IGF process? While sometimes it is reasonable to think that actors' sets are relatively bounded, in other cases the delimitation of groups to be studied is just an arbitrary choice (Wasserman and Faust 1994). Laumann, Marsden, and Prensky (1983) identified two modes of boundary identification: the nominalist and the realist approaches. Whereas the latter believes that boundaries are set up directly by actors according to their perceptions, the former approach is rather based on the concerns of the researcher and networks boundaries from conceptualizations of the object of study. Consequently, those who are considered from a nominalist point of view as belonging to a group do not necessarily perceive themselves as such, because we cannot assume that the perspective of the researcher corresponds totally to the reality (after all, it is a personal perspective based on individual understanding!).

For this study, we adopted a combination of the two approaches. First, a questionnaire was administered to members of those DCs that were launched between November 2006 and October 2007.[14] DCs were considered, from a nominalist point of view, as thematic and interaction proxies for the activity of mapping multiactor relations fostered by the IGF environment (see the section above "Dynamic Coalitions as Thematic and Interaction Proxies"). Starting from DCs implied, on the one hand, the possibility of narrowing our focus to "systematic participation" without equating *involvement* in the discourse formation solely with attendance at official meetings. On the other hand, this choice implied focusing attention only on a specific subset of particularly committed actors, with the risk of losing part of the broader picture. To avoid the risk of setting too narrow boundaries, each interviewee was asked to identify his or her partners in the broader IG domain, whether these were DC members or not and independently from their participation in the IGF process. We interviewed also those individuals who, although they were not initially included in the DCs member lists, were identified as partners by two different interviewees. In this way, we joined together a nominalist and a realist approach.

Once we concentrated on DCs, we also had to decide the level at which data would be gathered because Coalitions can host both individuals and organizations. A preliminary version of the questionnaire, which gathered data at both levels, was pretested in Geneva during the consultations on the way to the second IGF meeting (September 2007) and was then reduced to only the individual level. In fact, during the test phase many interviewees stated that they were affiliated with an organization but that they were bringing their personal opinions and perceptions. In this sense, we mapped individuals' relations and perceptions but, as it was not realistic to think that possible organizational affiliations were not influencing their positions (especially in the case of institutional or private corporation representatives), we considered status as an attribute of individuals. This allowed us to elaborate on multiactor dynamics beyond the individual level without forgetting that personal commitment is still a motor force in the renewal of IG mechanisms.

Despite their affiliation, interviewees were all DC members who participated at least once in the DC mailing list spaces from November 2006 to October 2007. If a mailing list was not employed in the DC communication strategy, we contacted the members listed on the DC website. The three repeated calls for participation were sent out using e-mails, while interviews were realized through different techniques: some exploited the Voice over Internet Protocol (VoIP) technology; some (the majority) were face-to-face interviews during the second IGF meeting in Rio; and some interviewees filled in the questionnaire on their own and sent it back though e-mail (after agreeing that, in case of doubts on the data provided, they could be contacted again). The joint employment of different methods for gathering data is justified by the difficulty of reaching DC participants. However, the questionnaire was administrated in all cases, and it is from that base that we obtained the data grounding our analysis.

Overall, we realized 49 interviews between September 2007 and June 2008 (36 percent response rate); 84 percent of interviewees were men; 71 percent of respondents were acting as representatives of non-governmental entitities (both the private sector and civil society); and 29 percent were representing institutions (both national and intergovernmental organizations).[15] Almost all respondents (46 out of 49) declared themselves to be involved in the IG domain with organizational affiliations, thus confirming the relevance of status issues also within open and participatory dynamics. Eighty-four percent of respondents believed that their jobs are actually related to the IG domain (they could be called "professionals"). As far as the level of individual experience in the field is concerned, the majority of respondents have engaged in IG matters during the WSIS process (55 percent). Thus, 71 percent participated in at least one of the WSIS meetings, but only a minority (26.5 percent) participated actively in the IG caucus during the Summit period.

As mentioned, the 49 interviewees were asked several questions from which we derived the data we used to design offline semantic and social networks. The networks we traced on the basis of the information retrieved in this way were analyzed to search for patterns both at the thematic and at the interaction level. We acknowledge that the number of individuals we reached is not representative of the vast crowd of IGF participants over the first mandate of the Forum as we are aware that we reached only a minority of DC members. Nonetheless, we consider our data as a useful entry point to explore general trends from the thematic and procedural points of view. In order to trace semantic networks, interviewees were required to describe their idea of IG. The themes mentioned, together with the answers, were linked in semantic networks that were then analyzed to uncover the dynamics of IG agenda enlargement and collective framing processes. Social networks were traced on the basis of partner-sharing relations that we derived from the identification of direct collaborators.

As far as online semantic and social networks are concerned, the data we employed refer to the period November 2006 to October 2007 and start from the two principal communication tools of almost all DCs: websites and mailing lists. To build up online thematic networks, the Issue Crawler[16] software was employed to draw several network mapping links between websites with the technique of the co-link analysis (see Chapter 4). The software traced online semantic networks using all URLs of organizations adhering to selected DCs that were listed online in January 2008. To trace online social networks, however, data were retrieved from eight DC mailing list archives in the period November 2006 to October 2007[17]—soon after the first IGF meeting in Athens and shortly before the Rio de Janeiro one. Online social networks are composed of those actors who have posted at least one message in the observation period and of ties existing between them and the message they posted (hence, they are two-mode networks; see Chapter 4).

In the following chapters we will illustrate the main results of the application of our analytic framework. We start from semantic and social networks traced online and then move to semantic and social networks offline.

NOTES

1. The historical reconstruction presented in the following sections draws heavily from the works by Hofmann (2006), Kleinwächter (2004, 2007, 2010), and Mueller (2002), who retraced the evolution of the IG across time up to the realization of the IGF. The reconstruction of the IGF process draws instead not only from academic and technical contributions but also from a systematic participant observation of the forum process conducted by the author in the period 2006–2008.

2. www.ietf.org/index.html.

3. From www.ietf.org/glossary.html#IESG.

4. www.iab.org. The IAB is responsible for defining the overall architecture of the Internet, providing guidance and broad direction to the IETF. The IAB also serves as the technology advisory group to the Internet Society and oversees a number of critical activities in support of the Internet.

5. At present, IANA is based at ICANN and still is in charge of "all *unique parameters* on the Internet, including IP (Internet Protocol) addresses" (www.ietf. org/glossary.html#I).

6. www.isoc.org/.

7. From www.ietf.org/glossary.html#I.

8. This work is not aimed at rediscussing the Domain Name System (DNS) matter and therefore will not indulge in detailed explanations on this point. So far, the "DNS war" (Paré 2003) has been the most explored IG matter from various perspectives, from regime theories to law-based approaches. A deep understanding of this problem and its ramifications requires extensive familiarity with the Internet structure and functioning system, with its governance arrangements, and with the legal, regulatory, and economic issues connected to management of domain names—such as

intellectual property rights regulations. For a deep analysis and a more precise reconstruction of the DNS controversy, see Mueller (2002) and Paré (2003).

9. www.wgig.org/index.html.

10. The list of WGIG members is available at www.wgig.org/members.html.

11. The WGIG final report is available in different languages at http://www.wgig.org/index.html.

12. Five major public policy areas are identified in the second part of the WGIG report as relevant for IG: (a) technical matters—domain names, IP addresses, rooting system connection costs; (b) spam, security, privacy; (c) commercial issues and intellectual property rights protection; (d) capacity building for developing countries; and (e) human rights: freedom of expression, consumers' rights, and multilingualism (WGIG 2005:5–8).

13. The first proposal is articulated in paragraphs 52–56 of the WGIG final report and foresees the creation of a Global Internet Council to replace ICANN GAC that would also perform some of the functions held by the U.S. Department of Commerce. The second proposal (para. 57–61) entails the enforcement of the GAC in order to meet the concerns of those countries asking for an enhanced governmental role. The third proposal (para. 62–67) is centered on the proposal to create an International Internet Council (IIC) that would assume ICANN and IANA functions and policy-drafting functions. Finally, the fourth proposal sees the restructuring of the IG system as revolving around three organizations: a Global Internet Policy Council (GIPC); a World Internet Corporation for Assigned Names and Numbers (WICANN), and a Global Internet Governance Forum.

14. The groups included in the survey were the Privacy Dynamic Coalition, Freedom of Expression and Freedom of the Media on the Internet, the Stop Spam Alliance, the Open Standards Dynamic Coalition, the Access to Knowledge@IGF Dynamic Coalition, the Internet Bill of Rights Dynamic Coalition, the Linguistic Diversity Dynamic Coalition, the Online Collaboration Dynamic Coalition, and Access and Connectivity for Remote, Rural and Dispersed Communities. As a consequence of the ongoing nature of the IGF process, some of these are now inactive, while some others, such as the Internet Bill of Rights Dynamic Coalition, merged with other groups that emerged lately and changed their name (in this case, the Internet Bill of Rights Dynamic Coalition became the Internet Rights and Principles Dynamic Coalition).

15. Very underrepresented between interviewees were firms and media (only one interview each). In both cases, this is justified by the scarce adherence these groups have to DCs but also to the explicit refusal by the private sector entities that were contacted to answer the questionnaire. The private sector individual and the media operatives reached were dropped into the nongovernmental category.

16. www.govcom.org and www.issuecralwer.net.

17. The only exception was the Stop Spam Alliance, which had no mailing list publicly available.

Chapter 4

The Internet Governance Online Discursive Space

During the Internet Governance Forum (IGF), remote participation online was guaranteed in both annual meetings and consultations (de la Chapelle 2010; Kleinwächter 2007, 2010). Moreover, IGF proceeding are currently published and are made publicly available on the IGF website, together with the proceedings of official meetings and the contributions sent by different actors during the whole process. The availability of all these documents is considered a very important step toward the enhancement of openness and transparency in the whole process (de la Chapelle 2010). Thus, a relevant portion of the multiactor dialogue happened online in between the official meetings and periodic consultations. In the context of IGF, the virtual space provided by the Internet becomes a further arena where multiactor dynamics can develop and the discourse on Internet governance (IG) can be renewed collectively.

However, acknowledgment of the relevance of online interactions and thematic confrontations has seldom been accompanied by systematic analyses of how the virtual space is inhabited by actors engaged in the effort of collectively constructing a discourse on IG. This gap is very much consistent with what emerged in Chapter 2, when we reviewed some of the reflections on the importance of the online space for political dynamics. In that context we saw that, in spite of the emphasis on the democratic potential of the Internet, investigations tended to search the online space for the *reproduction* of offline developed mechanisms or, in the best cases, to try to understand how online dynamics could favor the development of offline processes. In general, the online space is considered ancillary to the offline one, and its dynamics are analyzed *in comparison to* and not *in relation to* the offline world. Moreover, we pointed out that, while recognizing the complexity of the virtual world and the diversity characterizing different online communication tools, researchers have seldom studied online dynamics by considering more than one *modus communicandi* at a time (i.e., how online relations are established in different ways by passing from one communication device to another).

Starting from these premises, in this chapter we will attempt a structured analysis of the dynamics that are developed online to reduce the thematic and procedural uncertainty connected to the construction of an IG discourse. Consistent with our overall analytic framework, we will do so by exploring the semantic and social interactions developed online through the employment of two different online communication devices: on the one hand, websites and Web resources; on the other, mailing lists. What the two social and semantic networks have in common is that, as observed elsewhere (see Padovani and Pavan 2007, 2008), they grow up in a space that is less affected by constraints to active participation than the offline one. It is definitely true that there are some major obstacles to the consolidation of the virtual space as a public sphere (for example, the dramatic issue of the digital divide or problems connected to information overflows; see Papacharissi 2002 and Polat 2005) and that the Internet should rather be conceived as a space where multiple public spheres develop, interact, and overlap.

However, if we focus on what happens when actors do actually go online, it is undeniable that the online space offers multiple ways of participating in political processes and of generating content that can become a political resource. Moreover, equipping interested actors and stakeholders with the adequate infrastructure and the skills to go online and engage in the political dynamic implies more limited and sustainable costs than an actual financial strategy aimed at rebalancing the distribution of material resources that would be necessary to put all stakeholders on an equal footing. Consistent with this perspective, many efforts to promote online participation have been pursued within the IGF framework to exploit the potential of the online public space for hosting an unlimited number of actors and views, raising claims and articulating controversial issues, and providing actors with an actual chance to contribute to a multistakeholder process.

However, the openness and the ease of contributing to the discussion do not necessarily imply that online participation is universal or completely horizontal. Notwithstanding this fact, online patterns of inequality are seldom explored, thus leaving a hole in our understanding of how political dynamics evolve thanks to the employment of Information and Communication Technologies (ICTs). In an attempt to overcome the analytical shortcomings that derive from a general inattention to online dynamics in the IG domain, in the following sections we will analyze in the first place semantic networks in which different Web resources (websites in particular) are linked together.

This first level of analysis will allow us to uncover patterns along which a discourse on IG is being developed online through websites and their contents, as well as to articulate how linking strategies between websites impact the multiactor feature of the online IG conversation. Secondly, we will move to an examination of interpersonal communication through e-mails within mailing list spaces to explore the patterns through which online direct communication is structured and how it is sustained by more or less participatory conversations among mailing list members. Also, we will try to combine the structural features of online conversations with the topics discussed so as to obtain some insights on the very structure and purpose of online direct exchanges and to articulate the participative feature of the online dialogue on IG. In concluding this chapter, we will combine the results we obtained from the two levels of analysis to derive a general snapshot of how the virtual space is being exploited politically for experimenting with multiactor conversations on IG.

1. INTERNET GOVERNANCE ON THE WEB: ONLINE THEMATIC NETWORKS

The hyperlink logic upon which the World Wide Web is built is far from neutral and simple. In general, it is acknowledged that "there is a certain optionality in link-making. Making a link to another site, not making a link, or removing a link, may be viewed, sociologically or politically, as act of association, non association or dissociation, respectively" (Rogers 2008:4). The "selectivity" of the linking strategies pursued online helps to set up issues boundaries within the (theoretically) boundaryless virtual space: it is through links that actors are recognized, either as friends or foes, as belonging to that thematic territory or as strangers. In these last years of intensive growth of Internet communication and online content creation, several research projects have been developed to study Web relational structures, which represent practical evidence of this "optionality in link-making." In this context, "Web epistemology" (Rogers 2004) has been pushed forward as a general approach to the study of how information is circulated within the online space across different online communication devices and, in particular, along paths of links between Web resources focused on specific issues, that is, along issue networks.

> In terms of types of associations . . . issue networks might be distinguished from popular understandings of networks, and social networking, in that the individuals or organizations in the network neither need to be on the same side

of an issue, nor be acquainted with each other. . . . Actors may be antagonistic, oppositional, adversarial, unfriendly, estranged. Additionally, unlike social networks, issue networks do not privilege individuals and groups, as the networks also might be made up of a news story, a document, a leak, a database, an image or other such items, found on individual pages of Websites. (Rogers 2006:12)

Online issue networks well represent the complex collective articulation of a discourse within a potentially unlimited space where constraints to participation are lower than in the offline, real world (Padovani and Pavan 2008). As they gather online actors and resources dealing with a certain issue, online thematic networks *are based on* but *do not represent* the simple sum of links between a set of websites or online resources. Much more than that, they retrace the development status of online conversations carried on by those actors who are "speaking" online about a certain issue, whether they are adversaries or allies. The study of these online conversations can tell us first of all something about the *contents* of the discussion. When we look at what resources, websites, and documents are included in an issue network, we can elaborate on how that issue is treated and deployed online from a content point of view. Thus, the study of linking patterns allows us to elaborate on the prestige levels reached by actors within conversations as well as on the dynamics of exclusion from the conversation. Much like what happens in the offline world, inlinks (i.e., links that a Web resource receives from others) might be connected with the perceived relevance of a resource in the network structure, whereas outlinks (i.e., links that a Web resource sends to others) can be interpreted as the level of activity within the online network. Therefore, the study of online issue networks can also tell us something about how the process of discussion develops online.

We can therefore map and explore these structures resulting from a selective linking strategy so as to uncover how issues, concerns, and relevant matters are discussed online within interactional structures based on links and, in the end, to add a useful piece of knowledge on the overall processes of discussion (and confrontation) on them that are taking place offline. Indeed, online issue networks might represent *where the issue is based* within the virtual space in comparison to *where the issue is happening* in the offline world (Rogers 2004, 2006). These two locations do not necessarily coincide: the online space might host a different distribution of interests around an issue, such as IG, than the one taking place offline. And yet, these two locations are *intertwined* with one another, and one might wonder how much the offline development of an issue is reflected in the online and, vice versa, to what extent the online connections established among themes and actors online are then mirrored in offline political practices.

As mentioned above, the openness and ease of access do not necessarily imply that online participation is universal or completely horizontal. In fact, despite the potential to build online universal collaborations, within the Web space some identifiable structural dynamics are taking place and end up clustering some actors and contents together, in opposition to other groups and concerns or in isolation from the main online conversational patterns. Such dynamics are crucial for the online development of an issue and can actually mirror or influence how political conversations develop in the offline world.

When it comes to the IG case, we have seen that there is a plurality of spaces where the issue is actually "happening": some of these, such as the Internet Corporation for Assigned Names and Numbers (ICANN) or the Internet Engineering Task Force (IETF), are the spaces of the more traditional, technical IG; some others, like the IGF and the Dynamic Coalitions (DCs), are the spaces of broader discussions that are overcoming the purely technical horizon on this topic. It is undeniable that nowadays IG can no longer be conceived as a purely technical matter: many other issues, such as privacy, freedom of expression, and even climate change, have converged within its framework, especially with the realization of the IGF process. It is also quite clear that it is the convergence of different actors that is enlarging the agenda. We could therefore expect that the variety of places where the issue is happening offline would mirror in a crowded online issue networks where a vast variety of actors and themes are linked together. However, it remains an empirical question how much and in which way the real context of issue development is reflected online: Who is speaking with whom online about IG? What kind of discourse are actors helping to shape? Who should or could be there and is instead missing? How do we interpret, from a political point of view, the structural dynamics established within online issue networks? To answer these questions, we therefore map and explore what we have called in our framework online thematic networks (which correspond to the network type that is more suitable for exploring the semantic dimension of global communication within its online space) to master our understanding of how the IG discourse is developed though *conversational links* established among heterogeneous actors and themes in the vast and complex World Wide Web space.

1.1 Mapping Internet Governance Thematic Networks

Figure 4.1 shows the visualization of the online thematic network traced through the Issue Crawler, starting from the URLs listed on DC websites in January 2008.[1] The network represented in Figure 4.1 reproduces the main conversational paths established after two IGF meetings, at a time when IG

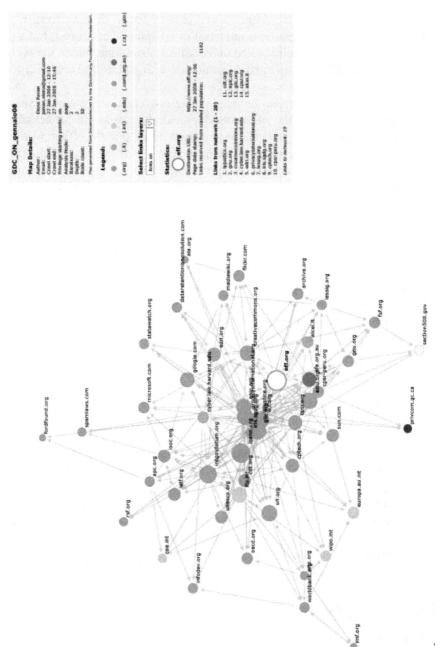

GDC_ON_gennaio05

Map Details:
Author: Elena Pavan
Email: pavan.elena@gmail.com
Crawl start: 27 Jan 2008 - 12:10
Crawl end: 27 Jan 2008 - 15:46
Privilege starting points: on
Analysis Mode: page
Depth: 2
Boundaries: 2
Node count: 50

Map generated from Issuecrawler.net by the Govcom.org Foundation, Amsterdam.

Legend:
(.org) (.it) (.int) (.edu) (.com/t.org.au) (.ca) (.gov)

Select links layers:
links on

Statistics:
eff.org
Destination URL: http://www.eff.org
Page date stamp: 27 Jan 2008 - 12:06
Links received from crawled population: 1182

Links from network (1 - 20)
1. ipjustice.org   11. cdt.org
2. gnu.org   12. epic.org
3. creativecommons.org   13. gilc.org
4. cyber.law.harvard.edu 14. cpsr.org
5. aclu.org   15. alcei.it
6. privacyinternational.org
7. lessig.org
8. iris.sgdg.org
9. cptech.org
10. cpsr-pecu.org

Links to network: 19

Figure 4.1.

had officially taken off as a global issue and the discussion began to be more structured as a consequence of the IGF process itself.

At a first glance, we cannot identify any clear-cut thematic patterns between the nodes. However, in the overall complexity of the structure under examination, we can notice some evident features of this network. Nodes are not equal among themselves: some of them have different sizes because they receive a different number of ties from others in the network; moreover, they do not all have the same color, and the different shades correspond to different top-level domains (TLDs) and, hence, to different types of websites. Finally, nodes are distributed within the space of the map; some of them are closer while some others are more distant. All these figurative elements help us visualize the different actors represented online through various Web resources, their different levels of prestige, and the clusters of nodes that share the similar relational patterns. Looking in more detail at our map of online relations, we can therefore wonder:

1. Who are the actors involved in the online thematic networks on IG? How diverse are they among themselves and, more importantly, in relation to the issues they bring into the conversation? In other words, what can we say about the openness and inclusiveness of the online discussion on IG?
2. What kinds of communicative relations are established between actors in the network? In other words, who is talking to whom and about what? What kinds of connections are built up between issues brought by actors in the network? Are there prevailing thematic clusters? Are thematic clusters connected or not?
3. How universal is the global discussion on IG in a space that provides fewer constraints to active participation? In other words, how many different political needs can we trace within our online thematic network? What are the missing elements?

The Online Diversity of Actors and Issues

Issue Crawler univocally identifies actors' types by the TLD at the end of their URLs and through a specific color. As Figure 4.1 shows, .org-domain nodes are the great majority in the network, confirming trends highlighted in previous analyses according to which there are many generic organizational and often noncommercial interests entering the IG online discussion (Padovani and Pavan 2007, 2008). However, some other types of actors also participate in the online discussion:

- .int for international organizations (the Council of Europe [coe.int]; the International Telecommunication Union [itu.int]; the European Union [europa.eu.int], and the World Intellectual Property Organization [wipo.int]);
- .edu for educational structures (the Berkman Center for Internet and Society at Harvard University [cyber.law.harvard.edu]);
- .ca for national Canadian websites (the Office of Privacy Commissioner of Canada [privcom.gc.ca]);
- .it for national Italian websites (Associazione per la Libertà della Comunicazione Elettronica Interattiva [alcei.it]);
- .gov for national governmental websites (dedicated to Section 508 of the U.S. Rehabilitation Act aimed at removing barriers in ICTs for disabled people [section508.gov]);
- .com for commercial activities (Microsoft [microsoft.com], Google [google. com], Sun Microsystem [sun.com], and the web space for picture sharing Flickr [flickr.com]); also associated with two noncommercial activities (the petition against data retention in Europe [dataretentionisnosolution.com] and a news portal on spam and security [spamlaws.com]).

The predominance of organizational entities is pretty much consistent with the overall and consolidated bottom-up approach to IG where non-institutional actors play a major role. Still, the overwhelming presence of .org nodes should not be interpreted as an indicator of continuity. In fact, as the .org domain tends to gather miscellaneous entities, innovation is detected if we look more closely at the components of the ".org population" in our network. Going under the surface of the TLD, we find that there are many different types of "organizations" involved in the discussion. In the first place, there are some relevant intergovernmental organizations, such as the United Nations (un.org), the World Bank (worldbank.org), the International Monetary Fund (imf.org), and the Organization for Economic Cooperation and Development (oecd. org). Also, we find IG-specific organizations, such as the International Engineering Task Force (ietf.org), the Internet Society (isoc.org), the World Wide Web Consortium (w3.org), ICANN (icann.org), and the Internet Governance Forum (intgovforum.org). Finally, there are also several groups that work in the area of digital communications and technologies yet with a much broader scope than the technical organizations we just mentioned: the Electronic Frontier Foundation (eff.org) and IP Justice (ipjustice.org), which are consolidated groups for the defense of civil liberties online; the Global Internet Liberty Campaign (gilc.org); the Association for Progressive Communication (apc. org), a well-consolidated civil society network working on ICTs and development issues; some groups particularly concerned with privacy issues, such as Privacy International (privacyinternational.org) and the Electronic Privacy

Information Centre (epic.org); some groups more broadly focused on digital rights (the European Digital Rights Initiative [edri.org]), civil liberties (Statewatch [statewatch.org] and the American Civil Liberties Union [aclu.org]), freedom of expression (Reporters Sans Frontières [rsf.org]), and intellectual property rights (creativecommons.org); and some others paying attention to the nested relationships between society, democracy, and ICTs, such as the Centre for Society and Technology (cst.org), or to the social impact of ICTs (Computer Professionals for Social Responsibility [cpsr.org]) and software liberties (the Gnu Foundation [gnu.org] and the Free Software Foundation [fsf.org]).

At a closer look, then, within our network there is much more variability in actors' nature than what the predominance of the .org domain would let us imagine. To uncover issue diversity, we have to look again under the surface of the TLD, which is only a superficial indication of actors' thematic concerns. Overall, different strands of thought seem to coexist within our network: beside the more traditional and technical perspective brought by traditional IG actors such as IETF and ICANN there is another, less technical perspective that is consolidating thanks to the online communicative activity of actors such as the Electronic Frontier Foundation (EFF), IP Justice, the Association for Progressive Communication (APC), Privacy International, Statewatch, Reporters Sans Frontières, and the American Civil Liberties Union (ACLU). These groups bring into the IG discourse a perspective based more on civil liberties than on infrastructure management, on the social rather than the technical aspects of IG. The predominance of nodes bringing a broader vision of IG is certainly a consequence of the fact that most DC adherents (from which we started) come from nontraditional IG forums. Therefore, the consolidation of a "humanistic" perspective on IG should not be overemphasized. And yet, in comparison with previous analyses (see Padovani and Pavan 2007, 2008), we can notice that the presence of nontechnical, publicly oriented initiatives within thematic networks has increased. The alternative, human-oriented perspective on IG has definitely become part of the mainstream discussion, but, overall, it has not replaced the more traditional and technical vision. Interestingly, as shown in Figure 4.1, both the technical and the humanistic perspectives on IG stand at the core of the network: ICANN and the W3C occupy very central positions in the representation, together with IP Justice, EFF, and Privacy International. Thus, the two strands parallel one another, creating an element of dynamism at the core of the IG online discourse.

Conversational Fluxes

Although we inferred from the diversity of nodes that the humanistic perspective and the more traditionally technical idea of IG are both in the core of our network, it is only by looking at relations within our network that we can elaborate on possible enmeshments between the two views. Has the humanistic perspective only scratched the surface of more traditional technical concerns, or has it definitely changed the way IG is conceived? Are the voices coming from IG traditional bodies searching for a dialogue with others from the civil liberties constituency?

In the first place, we can notice that while newcomers (i.e., civil liberties and public interest representatives online) do recognize through their links the traditional actors of the IG area (ICANN receives 14 ties from the network by miscellaneous actors, while IETF and the Internet Society [ISOC] receive 6), the reverse is not true. Traditional IG actors tend to talk only among themselves, or, in the best case, they link to institutional supranational actors such as the United Nations (UN), the World Bank, and the World Intellectual Property Organization (WIPO), who probably are considered as reference points by virtue of their regulatory activity. In between the technical and the human-oriented cluster there is the IGF website. As already pointed out in previous analyses (see Padovani and Pavan 2007, 2008), the IGF website keeps playing a joining role between the two perspectives on IG because it links to both traditional websites (e.g., ICANN, ISOC) and to newcomers (e.g., APC, IP Justice), thus connecting otherwise disjoined nodes. However, traditional IG bodies tend to recognize with different emphases the places where a broader IG discourse is carried on: while ICANN and ISOC do link to the IGF website, the IETF does not.

If we look at links sent by the public interest and civil society groups that are present in our network, we notice that the acknowledgment of IG traditional leading bodies is nevertheless less relevant than the internal communication flows established between actors concerned with broad social issues and civil liberties. In this sense, the possibility of building a truly open confrontation with more traditional actors seems limited by the predominant effort to foster a sort of internal coordination among very diverse social interests. However, in comparison with previous analyses, we can notice that the internal coordination of the heterogeneous universe of IG newcomers involves a more restricted set of themes: it entails mainly civil liberties (with special attention to privacy issues and digital rights), issues of software freedom, and matters pertaining to openness of standards.

Two other elements deserve some attention. In the first place, intergovernmental and worldwide organizations such as the WIPO, the IMF, or the World Bank are engaged in what could be called an online self-referential conversation. As noticed above, some non-institutional technical websites at the core of our network do recognize these supranational organizations, but, if we look at how these latter communicate with the rest of the thematic network members, we see that they only link among themselves, thus preventing any external access to a specific portion of the online conversation. Second, it is important to notice the nonconversational patterns enacted by (proper) commercial entities (i.e., Microsoft, Google, and Sun Microsystem): although they are recognized as partners within the discussion, they do not send any link to the rest of the network and constitute, in this sense, "network totems" (Rogers 2006, 2008).

How Universal Is Global Internet Governance Online?

Some might argue that asking questions about the transnational breadth of the online discussion on IG is pointless because, when we go online, territorial boundaries lose most of their importance. However, such a position presupposes an idea of transnationalism that is inherently connected with the nation-state and the offline real world. In fact, when speaking about transnationalism online, we should reframe the meaning of the term itself and conceive it in terms of the presence of local and regional political needs and perspectives, whether these are of a technical or social nature. We already noticed when we explored actors' diversity in terms of TLDs that the representation of local claims "is not an issue" in our thematic network. However, the very limited presence of local TLDs (.it, .ca, and .au) is somehow compensated for by the presence of local .org initiatives such as the Peruvian chapter of the Computer Professionals for Social Responsibility (CPSR [cpsr-peru.org]) and the American Library Association (ala.org). Finally, we can also retrace the European regional dimension, although only to a limited extent, thanks to the presence of the EU website and the petition against data retention within the European territory.

In general, these results confirm the trends highlighted in previous studies, according to which the online conversation on IG hosts only a limited representation of local realities (see Padovani and Pavan 2007, 2008). The absence of perspectives coming from the South of the world remains striking, especially if we consider that the IGF explicitly aims at enhancing the involvement of less developed countries in the political dynamic, thus bringing the Forum directly into places such as Brazil or India to limit the effect of physical participation constraints. In other words, if the Forum goes to the South, the online conversation remains based in the North. The only evident trace of a development-oriented discourse lies upon the link going from APC, the Berkman Centre for Internet and Society, and the Centre for Democracy and Technology (CDT) to the infodev.org, i.e., the global project that joins together some of the main international development agencies and that is hosted within the global ICT department of the World Bank. This particular link structure confirms an already highlighted trend according to which, whenever developmental issues have managed to enter the discourse, they have done so through the establishment of links going from civil society actors to institutional nodes (Padovani and Pavan 2008:11). However, institutions never reciprocate the links they receive. Therefore, it is reasonable to assume that, even though discourses related to ICTs for development are actually articulated, they hardly correspond to sustained and systematic strategies of action on the ground.

2. E-DEVELOPING A POLICY MATTER: ONLINE MAILING LIST NETWORKS ON INTERNET GOVERNANCE

If the analysis of online thematic networks helped us uncover the conversational dynamics that are sustaining the enlargement of the IG agenda through the construction of links between different Web resources, in this section we will try to identify if and how different thematic spaces promoted in the context of the IGF are hosting different interpersonal discourse development strategies. More particularly, we will examine the different use of e-mails to sustain specific activities within the online communicative spaces of DCs, that is to say, to sustain mobilization efforts in relation to specific IG subthemes.

We have mentioned in Chapter 3 that DCs are transnational voluntary groups that aim at developing specific issues belonging to the IG agenda and that work mainly in between the official IGF meetings to carry on dialogue between different stakeholders. The use of online communication tools therefore becomes central for the articulation of their issue-driven discourses, and, in particular, DCs rely heavily on the use of mailing lists. It is true that these mailing lists do not exhaust the whole of online interpersonal communications of the IGF and, even less, of the whole IG domain. For example, the Internet Governance Caucus (IGC; see Chapter 3) continues to carry on its activities through lively and sustained e-mail exchanges. Nonetheless, unlike all other online discussion spaces, DCs mailing list spaces were set up with the explicit intention of carrying on multiactor dialogue on specific themes pertaining to the IG framework when a physical presence is not possible.[2]

In general, one could argue that, if we look at public mailing list communications, we do not account for conversational dynamics where answers are sent privately to a restricted subgroup of mailing list members or solely to the message author. However, the choice of bringing the conversation offlist and transforming it into a private exchange inevitably alters the nature of the conversation, which ceases to be publicly available and a voluntary contribution to the collective shaping of the IG discourse. What is instead discussed within the public spaces of mailing lists can be thought of as evidence for voluntary engagement in public, multiple-actor conversational dynamics. It is in this sense that DCs mailing lists constitute public spaces for discussion that are created in response to the need for deepening specific controversial issues and that are potentially accessible by any interested individual or groups, not solely in connection with IGF annual meetings but at any moment in time. Hence, starting from the multiplication of DCs as an indicator of the progressive enrichment of the IG agenda (see Chapter 3), we can then study their use of e-mails to better understand the online interactional bases upon which the IG discourse is expanding.

As mentioned in the previous chapter, the DCs that have emerged over time do not have much in common. Beside a few shared features (being transnational, multistakeholder in principle, voluntary, and oriented to the deepening of a theme), DCs are rather heterogeneous groups that are free to organize as they prefer, and no clear-cut guidelines have ever been imposed. We can expect that this heterogeneity is evident first of all in the way they use their main communication tool, mailing lists. Our exploration of interpersonal online communication dynamics will be structured in a twofold manner. First, in order to search for regularities and differences in the use of e-mails, we will examine the *communicative volumes* of online exchanges in relation to some specific features of each thematic group (the size of the coalition, the number of individuals that have posted at least once in each coalition), and we will compare the longitudinal variations in the use of e-mails to the roadmap of the IGF process. Second, exchanges within each DC's communicative space will be translated into networks, with the overall aim of analyzing what kinds of interactional dynamics are actually set up within each group to sustain the development of a discourse around specific IG subthemes.

2.1 Online Communicative Volumes

As Table 4.1 shows, Coalitions' communicative spaces host a different quantitative use of e-mails,[3] namely, diverse communicative volumes. Overall, the Privacy group makes the most intensive use of mails (see its overall communicative volume of 192 messages in 12 months). All other Coalitions we mapped use mailing lists to a lesser extent, but, starting from their exchange volumes, we can identify groups: on the one side, the Open Standards, the Freedom of Expression, and the Online Collaboration DCs, which use their mailing lists rather similarly (10–12 message per month on average); and, on the other side, the Internet Bill of Rights and the Access to Knowledge

Table 4.1. Dynamic coalition mailing list activity rates, November 2006–October 2007

Dynamic Coalition	Number of Posts													Size	Active Posters
	11 06	*12 06*	*1 07*	*2 07*	*3 07*	*4 07*	*5 07*	*6 07*	*7 07*	*8 07*	*9 07*	*10 07*	*Tot.*		
Privacy	5	5	26	34	11	8	41	24	9	6	13	10	192	69	37
OS	19	4	16	21	11	23	9	8	10	3	7	24	155	14	25
FoE	10	0	5	8	6	34	30	21	4	1	10	18	147	15	34
OC	0	0	5	50	6	2	19	1	0	0	10	27	120	23	16
IBR	4	0	4	17	11	4	12	9	7	4	15	14	101	11	17
A2K	14	1	9	15	4	22	18	3	9	1	3	1	100	20	24
LD	0	0	0	0	0	0	0	0	0	5	5	15	25	5	9
ACRDC	0	0	0	0	0	0	0	3	0	0	7	0	10	—	6

Note: OS = Open Standards; FoE = Freedom of Expression; OC = Online Collaboration; IBR = Internet Bill of Rights; A2K = Access to Knowledge; LD = Linguistic Diversity; ACRDC = Access and Connection for Rural and Dispersed Communities; — = data not available.

coalitions, which use their mailing lists less often than the Privacy DC (8 messages per month on average).

Overall communication rates are, in some cases, consistent with the size of the various coalitions as measured through the number of supporters listed on the websites[4]: the Privacy DC has the larger number of adherents and is also the group with the highest mailing list (ML) communicative volume, whereas the Internet Bill of Rights is much smaller and its ML volume is, consistently, more limited. Nonetheless, the relation between size and online activity is not valid for all cases. Indeed, the Freedom of Expression and the Open Standard DCs use e-mails quite intensely despite the rather limited size of these coalitions. Looking at the number of active posters within each group (i.e., single individuals posting at least once during the observation period), we find that only in some cases does a higher number of posters correspond with a higher volume of communication online (e.g., in the case of the Privacy DC). In fact, there are exceptions: the A2K group has a rather high number of posters (24) but a limited use of its ML; conversely, the Online Collaboration DC has a lower number of active posters (16) in spite of a relatively high volume of exchanges. In the end, there seems to be no relation between coalitions' group features and their employment of e-mail to foster communication between members.

If we move to observe how each coalition's e-mail exchanges evolve longitudinally, we see that no ML shows a constant growth and, therefore, ML communication remains somehow functional to the pursuit of specific tasks. Periods of more intense activity are followed by periods of lower intensity or even by the total absence of communication between members (as happens in December 2006 for the Freedom of Expression and the Internet Bill of Rights). Moreover, all groups (with the exception of the LD and the ACRDC) show particularly high activity rates in the period that from November 2006 to May 2007 and then cool down, hardly surprising, over summertime and become active again near the second IGF meeting (November 2007). In general, we notice that exchanges within coalitions tend to grow in relation to IGF consultations (February, May, and September 2007), while communications tend to slow down during the following month (the sole exception being the last consultation before the Rio meeting, which was immediately preceding the official event and, therefore, the pursuit of further organizational tasks kept conversations alive). At first glance, then, it seems that online exchanges vary more according to the development of the IGF offline process rather than as a consequence of coalitions' attributes.

However, the offline deployment of the IGF does not affect online exchanges of coalitions to the same extent. When approaching the realization of an official Forum meeting, communication exchanges in all groups tend to increase, but, all in all, this finding is hardly surprising: after they have received institutional recognition from the IGF Secretariat, Coalitions have been allotted ad hoc slots within the event program, and their representatives are allowed to speak up on behalf of the group during the consultation. In this sense, all groups have to face internal organizational matters, and, to this end, they exploit e-mails to organize their event in the IGF days or during their contribution to the consultation. Still, only in some cases, as for the Privacy Coalition and the OC Coalition, the change of communication volume seems to be more precisely determined by the roadmap of the IGF process. For some other groups—the Open Standards Coalition, the A2K Coalition, and the FoE Coalition—communication peaks instead of falls outside the months of open consultations, and, therefore, it seems that the construction of discourses within them is following other paths that are not necessarily paralleling the offline process.

Summing up, if we look at e-mail exchanges from a general perspective based on coalitions' attributes, no clear pattern seems to emerge, and the influence of organizational matters in preparation for official events seems limited to the stimulation of conversational trends but still far from being fully determining. In the end, though, we cannot be that surprised by this finding, as the absolute absence of guidelines on DCs formation and their functioning allows very discretional organizational and communicative choices. However, if we cannot find that many regularities on *how much* the

IG discourse is developed online, we might wonder if there are similarities on *how* DCs structure their online conversations and see if there are commonalities between groups based on the types of interaction they host. In the end, multiactor dynamics can be sustained and highly participated in even though not so many e-mails are sent, but conversations are participated in and happen in a sustained way instead of being episodic and pointing to fragmented interactions. To investigate how multiactor dialogue is carried on online, then, we should move from the examination of *communicative volumes* to the study of *communicative dynamics* between actors online.

2.2 Online Communicative Dynamics

To examine what kind of communication is taking place within groups involved in the articulation of specific aspects of the IG domain, it is necessary to shift from the main features of communicative exchanges to the type of relations that are established among actors online. A first step in this sense is provided by the analysis of threads discussed within each communicative space. Table 4.2 presents the overall number of threads identified for each ML,[5] the maximum number of individuals involved within one thread[6] (i.e., what we call the "maximum popularity" reached by discussion within a group), the total number of posts sent within each group, and the number of active posters. As shown, the Open Standards DC is the group within which the highest number of different threads has been proposed, followed by the Privacy group. Internal discussion within other groups was instead sustained by a more limited number of posts. Moreover, comparing the maximum popularity of threads, it seems that discussions developing within the OS group and the FoE group are those reaching out for a broader general participation. Therefore, at first glance, we can infer that there are some differences among DCs in how online conversations are developed, both in terms of the

Table 4.2. Number and popularity of threads for each Dynamic Coalition (DC)

	Privacy DC	OS DC	FoE DC	OC DC	IBR DC	A2K DC
Number of threads	79	95	47	40	42	55
Maximum popularity	9	10	12	5	7	8
Total number of posts	192	155	147	120	101	100
Active posters	37	25	34	16	17	24

Note: OS = Open Standards; FoE = Freedom of Expression; OC = Online Collaboration; IBR = Internet Bill of Rights; A2K = Access to Knowledge.

articulation of discussion (i.e., number of threads) and in terms of participation potential (i.e., maximum popularity of a thread).

However, the fact that groups differ in the number of threads or that some threads attract more participation than others is just one aspect of the general participation to online discussions. It might be interesting to also explore how exchanges around single threads develop, in terms of how many posters contribute to the development of online matters. To study this point, eight different two-mode networks were reconstructed starting from ML archives. As all two-mode networks, these networks gather two different sets of nodes (Wasserman and Faust 1994), which, in our case, are active posters and threads initiated in the ML space. Thus, there is a tie between one node belonging to the first set and one node belonging to the second set if active posters contributed to a certain thread. Overall, the resulting two-mode networks represent a snapshot of one year of communication among active posters. Figure 4.2 shows an example of the two-mode network we built in this way: circles represent individuals sending e-mails while squares represent the threads to which individuals have contributed through their messages. There is a tie between a circle (one individual) and a square (a thread) if that individual contributes once or more to the development of the discussion in relation to that thread. If one individual writes several e-mails concerning different threads, her circle links to several squares. Moreover, as shown in the figure, the thickness of the ties is proportional to the number of e-mails one individual sent in relation to the same thread of discussion. Hence, the bolder the tie, the higher the number of messages sent to that thread by the same person and, therefore, the stronger her contribution in the discussion of topics.

To study in more detail the conversational dynamics from these bases, we can analyze regularities at the level of subgraphs (Wasserman and Faust 1994:506); that is to say, we can analyze how conversations online are structured around minimal compositional elements. Simmel (1998 [1908]) and his study of the consequences of group size on social processes provide a useful entry point to this aim. According to his approach, the smallest and most unstable form of group is provided by the *dyad* or, in other words, the union of two previously separated individuals by means of a tie. The dyad is the minimal form of group because only isolated individuals precede it; but it is also unstable, because when one of the two individuals withdraws from the couple the group ceases to exist. If a third node is added to the pair of actors in a dyad, the group becomes a *triad*. This triple set of actors and the ties between them allow for the formation of internal coalitions (a majority of two against a minority of the remaining individual) and allow the internal relational structure to expand from intimacy (characterizing dyadic exclusive

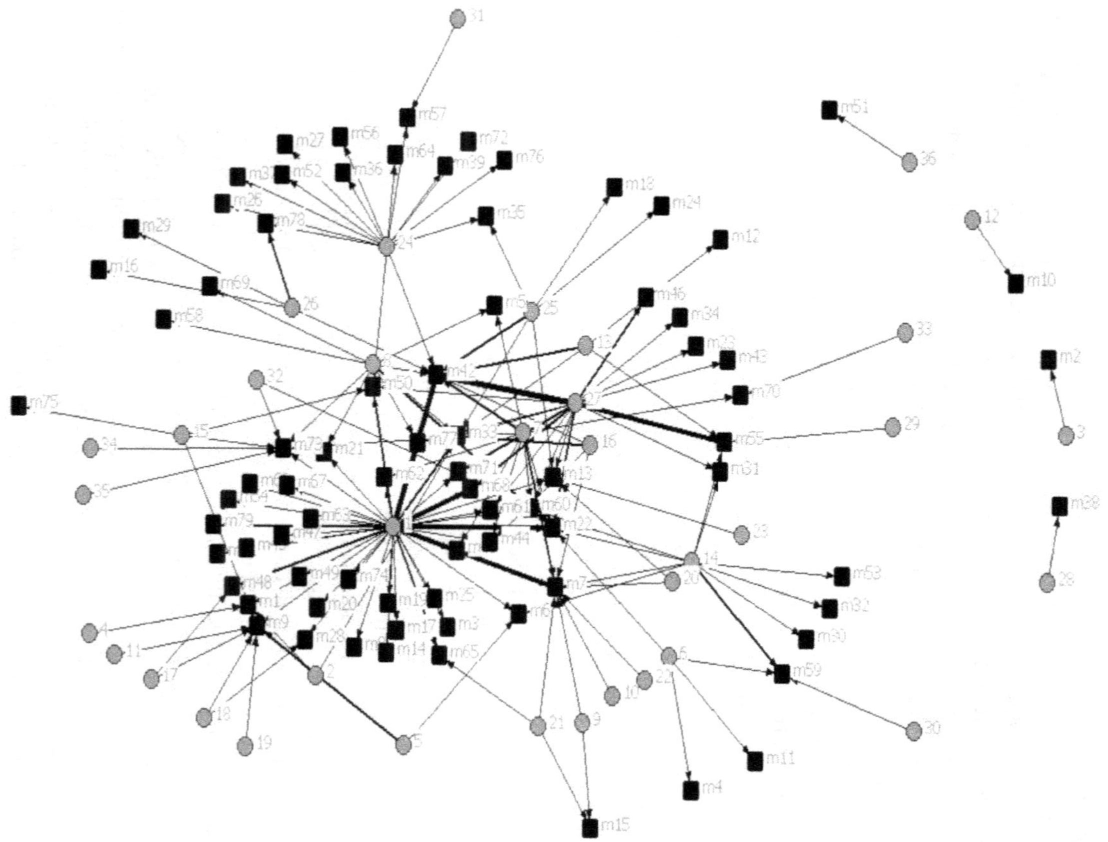

Figure 4.2.

Table 4.3. Types of threads within dynamic coalition mailing lists

	Privacy DC	OS DC	FoE DC	OC DC	IBR DC	A2K DC
Informative threads	60.8%	81.0%	59.6%	40.0%	57.1%	74.5%
Dyadic threads	20.3%	10.5%	12.8%	25.0%	21.4%	11.0%
Triadic threads	6.3%	2.0%	8.5%	17.5%	9.5%	7.3%
Larger threads	12.6%	6.5%	19.1%	17.5%	12.0%	7.2%
Total	100%	100%	100%	100%	100%	100%

Note: OS = Open Standards; FoE = Freedom of Expression; OC = Online Collaboration; IBR = Internet Bill of Rights; A2K = Access to Knowledge.

relations) to a most heterogeneous set of arrangements. Finally, larger groups can be read in terms of how they combine dyadic and triadic arrangements.

Starting from the original ideas of dyads and triads established between social actors, we can analyze exchanges within mailing lists in search of how conversational groups are shaping in more stable or unstable ways. In our case, as groups are formed around threads along conversational dynamics, we apply the concepts of dyads, triads, and larger groups in a slightly revised form, so that they embody the participation of two, three, or more individuals in articulating threads. It is reasonable to assume that the higher the number of individuals involved in discussion the higher the level of participation and the more complex the articulation of IG discourses. Table 4.3 shows, for every DC we considered, the number of *informative threads*, consisting of messages sent by one individual to which none replied[7]; *dyadic threads*, conversations on the same thread between two individuals; *triadic threads,* involving three individuals; and, finally, *larger threads* occurring in groups of more than three individuals. As shown in Table 4.3, the majority of threads within all groups belong to the first category (i.e., informative). Online communication dynamics are then quite fragmented, as most e-mails sent within DC mailing lists do not entail proper "exchanges" but, rather, inputs to other members.

What makes the differences between groups is the number of other types of communicative dynamics. In this sense, besides the predominance of *informational threads*, we notice that the Privacy DC and the IBR group see a prevalence of *dyadic threads* over other types of exchanges. The Open Standards group is sustained instead by very restricted exchanges: a dramatic predominance of *informative threads* is accompanied by a small number of *dyadic* exchanges and by the residual presence of more participatory dynamics. Similarly, although with different proportions than those of the OS coalition, the A2K group does not seem to host large discussions among its members. Finally, the Freedom of Expression and the Online Collaboration DCs tend to counterbalance their predominant *informative* feature of communication with a rather high presence of *larger threads* involving a higher number of individuals. Overall, then, different thematic strands of IG discourse, represented by the main thematic focus of DCs, are built up in a rather fragmented way because most ML exchanges translate into unilateral submissions of discursive inputs and are sustained by minimal and unstable groups (the dyad). Larger interactional structures and, therefore, the possibility of debating at large IG matters are still less diffused although not totally absent.

But is the thematic focus of DCs the very subject of online threads that have large participation? In this sense, we could wonder about the *participatory substance* of each DC, namely, about the topic that is discussed within more participatory threads. Crossing communicative patterns (i.e., informative, dyadic, and larger) with the topics of more participatory threads, we notice an interesting consistency among groups (see Table 4.4):

1. Predominantly informative groups such as the Open Standards and the Access to Knowledge groups experiment with higher participatory conversations when it comes to matters of internal organization, such as the draft of a coalition process document (in the OS case) or of an internal meeting outside the IGF time schedule (in the A2K case). In other words, the most participative threads within informative groups are instrumental to the group itself and to its organization, while more issue-oriented threads tend to be less participative and take the form of information-sharing exchanges.

2. Predominantly dyadic groups such the Privacy and the IBR DCs have higher participation when it comes to organizing autonomous initiatives significantly related to their main thematic area but *outside* the IGF space. In particular, for the Privacy group the Montreal Conference on Data Protection and Privacy Commissioners[8] functioned as a stimulus for the production of a set of papers and documents to collectively address

Table 4.4. Classification of Dynamic Coalitions based on the combination of communicative patterns and more discussed topics

		Topics of highly participated threads	Type of group
	Informative	Internal organizational matters	Informative group, A2K and OS
Communicative Pattern	Dyadic	Organization of events outside the IGF	Dyadic groups, Privacy and IBR
	Larger	Focus on their very thematic interests	Discursive group, FoE and OC

complex matters such as identity management, while for the IBR group the organization of the first edition of the *Dialogue Forum on Internet Rights*[9] represented a further effort paralleling the engagement in the IGF process. In other words, the most participative threads within dyadic groups supplement the high information-sharing level with dynamics to manage organizational tasks to promote a discourse on a specific IG sub-theme but outside the IGF milieu. While the majority of issue-oriented topics are discussed through dynamics with low participation, larger participation is stimulated in relation to the attempts to consolidate thematic concerns within external initiatives.

3. Finally, in the case of more discursive groups (the Online Collaboration and the Freedom of Expression DCs), more participative threads are very much inherent to the thematic focus of the coalition. In the case of the OC group, the most animated discussion pertained to the amelioration of remote participation during and in between the next IGF meetings. In the case of the FoE group, the more animated discussion revolved around the .xxx controversy and ICANN's regulatory activity on it. In this sense, these two groups stimulated more participative exchanges on their topics of interest, feeding communicative and inclusive discussions and paralleling, in this way, the presence of organizational threads with a general dialogic and issue-oriented activity.

3. DISCUSSING INTERNET GOVERNANCE ONLINE: A (NOT SO) GLOBAL AND PARTICIPATIVE CONVERSATION

This chapter started by acknowledging the inherent relevance of the online dimension for a correct and complete understanding of the current phase of discussion on IG. Thus, it moved from a twofold critique to existing analyses of online political dynamics. In the first place, although contemporary literature on the relationship between the Internet and political behaviors has underlined the complexity and the nonlinearity of the relation between these two spaces of interaction, it has also considered the online dimension not as a space per se but rather as an ancillary or an extension of the offline world. This attitude led to an understanding of online political behaviors in *comparison* to and not in *relation* to the offline world and jeopardized an accurate examination of how online political relations can structure and develop autonomously, thus relating to (and, possibly, influencing) offline dynamics.

Secondly, we argued that, if our aim is to overcome these limitations and to achieve a genuine understanding of the online dimension of current political dynamics in the IG domain, it is necessary to consider how IG is treated, managed, and developed through different online communication devices. Only a joint interpretation of results coming from a multifaceted analysis can provide us with a more comprehensive vision of what the online space really means in the IG case. Starting from these premises, in this chapter we examined IG online discursive dynamics as they develop in the virtual space both from a thematic and a procedural perspective, using, respectively, online issue networks and mailing list exchanges. In concluding this chapter, we summarize and put together the results we obtained by the analysis of online semantic and social networks to get a snapshot of how the virtual space is shaping, in the IG case, as a political space for the development of multiactor conversations.

In general, the analysis of online thematic networks showed us that a technical and a more humanistic vision of IG (based, respectively, on Internet critical resources management and on civil liberties, privacy, openness of standards, and software) are both present within the core of the discussion. Thus, neither seems to prevail over the other: the most prominent actors in the online thematic network indeed represent both positions (ICANN for the traditional one and EFF for the more socially oriented one). This finding is particularly interesting because it suggests that, after a few years of discussion fostered by the realization of international participatory processes such as the IGF, online conversations over IG seem to have formed around two main and equally relevant strands and that, therefore, the IG is more than just cables, domain names, and IP addresses.

Nevertheless, at a closer look, we noticed that what is developing in the online space is not so much an *enmeshment* between the two perspectives but, rather, *the consolidation of two parallel discussions* whose interplay is still limited. Indeed, technical matters continue to be developed among the usual IG bodies, who largely ignore the presence of newcomers concerned with the social implications of IG and only in some cases reach out to intergovernmental institutions by virtue of their regulatory function. Consistently, the more recent socially oriented perspective on IG is evolving quite autonomously around specific concerns such as privacy and civil liberties, thus seldom recognizing the "authority" traditionally associated with older IG organizations. When this recognition between new and older IG protagonists actually happens, it is not reciprocal, and the traditional, technical discourse on IG does not seem to enmesh with socially oriented matters. In sum, then, if the IG agenda has changed over time (and this is a remarkable element), we are at a point where different visions of IG are juxtaposed and not fully integrated. In this context, the IGF provides a space where the two perspectives can meet. In the online thematic network, the official process works as a juncture to coordinate the simultaneous presence of traditional and more recent actors and themes in the domain.

The study of online thematic networks pointed out three other critical trends. In the first place, it showed the self-referential feature of conversational fluxes between intergovernmental nodes and international organizations, who, although they are sometimes called into question by nongovernmental IG actors, never reciprocate these links and relate only among themselves. Secondly, it showed the nonconversational behavior adopted by the private sector, which is behaving like a "network totem" (Rogers 2006) that does not reciprocate the attention other nodes give to it. Finally, despite the premises of easier access and of increased participation thanks to the exploitation of the virtual space, the online discursive milieu appears mainly concerned with global and overarching issues, leaving localisms and specific claims outside the discussion. Whenever developmental issues are addressed, the relation is still unidirectional and goes from civil society to institutional actors with no way back. In this sense, it appears that, if developmental claims are raised in the IG domain, governmental actors are still finding it difficult to reply to them.

Moving to the study of conversational dynamics within multiactor spaces, the examination of mailing list exchanges showed that there is not much homogeneity in the quantitative use of e-mails made by DCs. This result is hardly surprising given the absolute absence of guidelines on the formation and functioning mechanisms of Coalitions. Each group decides at its own discretion and on the basis of available resources how to internally organize and communicate via e-mail. However, if we cannot find systematic regularity on *how much* the IG discourse is developed online within groups, we actually find that there are similarities on *how* DCs structure their online conversations. The examination of relations sustaining the articulation of online discourses within each DC showed that communications are rather fragmented and seldom reach out to a larger group level.

Looking at how exchanges of e-mails are structured by minimal relational structures (dyads, triads), we noticed that most e-mail communication online is of an informative nature and is not sustained by proper group dynamics: DCs mailing list participants send single messages without generating any public feedback from the rest of the group. This communication dynamic is common to all DCs, but it is sustaining in particular the online discourse on open standards and access to knowledge in which many inputs are provided but no real conversation develops. Only in the cases of FoE and OC Coalitions do we find larger discussions that are able to counterbalance the dramatic predominance of online communicative fragmentation. Furthermore, in looking at the actual topics discussed in more participative threads, we noticed that most of the time organizational matters stimulated greater participation, while only in the cases of FoE and OC Coalitions did highly participative threads pertain to the thematic focus of the group. In sum, the consolidation of participatory dynamics with a specific thematic focus seems still limited, and, in our study, it characterizes the discourses on multistakeholderism enhancement and on freedom of expression. Although internal organization tasks can still contribute to the development of a discourse on IG (e.g., when DCs members organize internally to present a common position during consultations), the use of e-mails to actually widen the IG agenda, to exchange opinions and viewpoints, and to resolve symbolic conflictual challenges seems still limited.

What is the overall picture that emerges from this twofold investigation of the IG online space of discussion? The online IG discursive space is heterogeneous and composed of a plurality of actors and themes connected among themselves, as well as by a plurality of relational modes of discussion. Both issue networks and social networks are inhabited by governmental and nongovernmental actors; since both online issue networks and social networks show that a multiplicity of themes are being discussed in relation to the overarching idea of IG, the online agenda results are very variegated. From the point of view of semantics, analysis of the online thematic networks showed that the more traditional, technically oriented IG discourse is paralleled by a more recent, socially oriented position on IG matters. From the point of view of social relations, analysis of DCs mailing lists showed that the convergence of different voices in the same space is sustained mainly by the submission of thematic inputs to the groups while larger conversations are less frequent.

More generally, both levels of analysis revealed the inherent complexity of thematic and procedural interactions in the online IG space of discussion. Thus, the examination of online social and semantic networks invites us to question the actual multistakeholder feature of online exchanges as well as the actual deployment of an IG *discourse* in terms of participated exchanges and confrontations among positions. Indeed, the possibility offered by the Internet to feed and articulate different perspectives on the same issues as well as to communicate more easily with others seems not to be fully taken as a fruitful point of departure to construct multithematic and multiactor relations. In particular, the results of our study of online thematic networks show us that thematic exchanges are established between civil society entities, whether they are technical representatives or more concerned about the social implications of IG. If we look at relations established by intergovernmental organizations and the private sector online (and not simply at *how much* they are represented online), we notice that the very limited contribution they provide to the overall discourse online does not correspond with their actual participation in the offline process. As observed elsewhere (see Padovani and Pavan 2008), their conversations must be carried on somewhere else, perhaps also online but neither in connection with nor exploiting spaces for multiactor dialogues fostered by the IGF. Overall, then, conversations are carried on separately even online, where the construction of a link would be much easier than building a tie offline.

Furthermore, the analysis of mailing list networks showed that the online articulation of IG discourses fostered by the IGF process itself is carried on in a rather fragmented way and only seldom entails the formation of large conversational groups. Large e-mail exchanges sustain only very rarely an actual thematically focused discourse on the *substance* of IG matters. More often, larger discussions deal with internal organizational tasks. This result seems very much consistent with the trend emerging from the examination of the online thematic network, where a push for internal consolidation overwhelms the construction of links between groups and interpretations. In sum, the online convergence of perspectives is, in the end, still limited and tends to be pursued internally, rather than across groups.

It could be argued that the online low profile of institutions and of the private sector is a bias produced by our methodological choice of mapping activities from DCs and that the fragmentation of personal communication is a bias inherently connected with the choice of studying mailing lists. In fact, here we are not suggesting that the problem consists in the low level of representation reached by institutional and business actors online or that the private dimension of communication is overwhelming the public one. What we are arguing here, instead, is that the critical points are the noncommunicative behavior shown even by the few institutional and business actors that are present online and the overall fragmentation of communication exchanges that should contribute to uncovering the complexity of IG matters.

In concluding this chapter, then, we could summarize the results obtained by our two levels of analysis by saying that the online IG discursive space appears heterogeneous in terms of themes but also fragmented and rather exclusive from a process point of view. The lack of constraints for including or discussing topics online has certainly contributed to the enlargement of the IG agenda and has increased the potential for the construction of conversational dynamics joining together actors in a multiactor discussion. In this sense, the vitality of the IG discursive space is one of the major features of contemporary political dynamics in this domain. However, the open character of the online space has not fully translated yet into a global and participatory multiactor conversation. Rather, it has brought a proliferation of multiple strands of reflections, carried on mainly by technically and socially oriented civil society members that engage in the IGF process while governmental representatives and business entities prefer to discuss their positions separately. A genuine multiactor conversation, where also governmental and private sector entities participate, appears quite far away. Also, the enmeshment of the two main perspectives supported by civil society members seems jeopardized by the low level of mutual recognition. The low level of actual direct conversation going on online further complicates the overall situation. Even if discussions online have contributed to the proliferation of themes and inputs, the openness of the online world has not automatically translated into inclusiveness, either at the thematic level, where specific claims do not find a place, or at the interpersonal level, where multiactor groups fail to take shape beyond the unilateral submission of inputs and minimal conversations. Quite ironically, the online discussion on global IG seems far from being global itself.

NOTES

1. The Issue Crawler working logic for producing thematic networks is presented in the Appendix.

2. As our focus in this work is set on the dynamics fostered by the creation of the IGF, we did not use mailing list archives of the IG Caucus (IGC) because its creation dates back to the World Summit on the Information Society and, in this sense, it represents an important locus of discussion for civil society individuals and entities. Whether after more than 5 years of activity the IGC mailing list has ended up by informally including also institutional members (for some individuals might have changed affiliation or might have become members of governmental task forces on IG) or not, it was born as a place for civil society discussion and, hence, it lacks the multiactor feature that characterizes (at least in principle) DC spaces.

3. The Linguistic Diversity group and the Access and Connection for Rural and Dispersed Communities group were established later than the other groups and, for this reason, they are not included in the evaluation of communicative volumes.

4. It is important to remember that, because DCs do not have clear-cut membership criteria, their size can be evaluated only indirectly. In this case, the number of declared website supports registered on each coalition website was employed as proxy for the size.

5. The Linguistic Diversity and the Access and Connection for Rural and Dispersed Communities groups have not been considered here, as their activity has been too narrow within the observation period.

6. Because originally matrixes were weighed according to the number of messages one individual had sent in relation to a specific thread, in order to count the number of individuals involved within one thread and, hence, the maximum thread popularity for each ML, matrixes have been dichotomized.

7. In terms of social network analysis, *informative threads* can be equated with null dyads, in which "neither actor has a tie to the other" (Wasserman and Faust 1994:511).

8. Montreal, Canada, September 25–28, 2007.

9. Rome, September 27, 2007.

Chapter 5

The Internet Governance Offline
Discursive Space

In 2005, participants to the World Summit on the Information Society (WSIS) concluded a 5-year effort by writing in the Tunis Agenda: "We ask the UN Secretary-General, in an open and inclusive process, to convene, by the second quarter of 2006, a meeting of the new forum for multi-stakeholder policy dialogue called the Internet Governance Forum (IGF)" (art. 72). Since then, comments on the value of the IGF process for consolidating multiactor communication have multiplied, thus stressing the relevance of this institutional experiment not only for the Internet governance (IG) domain but also, more broadly, for the reform of supranational politics toward an enhanced participatory and democratic model.

Systematic analyses of how this policy dialogue is forming are still missing. It has not been analyzed yet how multiactor cross-participation influences the definition of IG thematic boundaries or how political relations are being structured between the different stakeholder groups. It is unquestionable that the IGF is fostering the progressive enlargement and the systematic organization of a broad IG agenda, as it is acknowledged that, over time, governments, intergovernmental organizations, business entities, public interest organizations, and the technical community have found in this forum a space where they can confront one another and derive lessons to be brought "back home." As stressed by Kummer (2010), the IGF actually works as an "incubator" of ideas, but so far there have been few attempts to analyze how this "incubation" process is developing. An exploration of the dynamics that are operating offline at the thematic and the procedural levels therefore becomes crucial. In the first place, it can provide insights on how opportunities derived from the establishment of an institutional occasion to realize a policy dialogue are actually seized to reduce IG twofold uncertainty. Secondly, it allows the identification of some crucial nodes in relation to broader reflections on how meaningful precedents in experimenting with multistakeholder processes are being created in the Global Communication Governance (GCG) field.

In the following sections we will then explore how IG contents and processes are being defined thanks to the joined participation of actors in the same policy dialogue process. At the thematic level, we will first explore how different understandings of IG can be organized in terms of the cognitive dimensions underlying individual frames, what we call *orientations* in the construction of the IG discourse. Moreover, semantic networks will be employed to analyze how different thematic inputs brought by stakeholders within the discussion are linking to one another and are progressively structuring the discourse on IG. Indeed, although all claims and instances are welcomed within the discursive space provided by the IGF, their level of integration within the IG discourse varies: some themes become more prominent, whereas some others occupy more peripheral positions or work as a link between the more technical and the more socially oriented perspectives. At the process level, instead, we will explore though social networks what kinds of dynamics are underpinning multiactor conversations stimulated by the IGF. In this sense, we will try to identify if there are groups that host more intense conversations and investigate what individual elements can influence the communication potential held by different actors in the offline collaboration network that is fostered by the IGF process.

1. WHAT ARE THEY NETWORKING ABOUT? OFFLINE SEMANTIC NETWORKS ON INTERNET GOVERNANCE

In this section we will explore adopting a relational perspective on how the thematic boundaries of the IG discourse are being progressively defined. To do so, we will start from answers provided by 48 of our interviewees[1] to the questionnaire item asking them to associate up to three themes with the label *Internet governance*. As shown in Table 5.1, themes identified by respondents in their answers shape an enlarged IG agenda that goes well beyond technical issues (although these are still central to the discussion; see, for example, Internet critical resources, ICRs). Indeed, a set of new issues such as security, access, and human rights appear as increasingly relevant, sometimes overcoming more traditional issues in the number of mentions provided by the respondents.

However, given the open and participatory features of IGF, the increasing variety of issues linked to the IG label is hardly surprising: participants to the process bring to the discussion table their own claims and concerns, and these, in turn, legitimately enter the overall IG discourse. Looking at the list of themes

Table 5.1. Thematic inputs on Internet governance

Theme	Frequency	Theme	Frequency
Security	20	Content regulation	1
Access	19	Deliberation	1
Internet critical resources	17	Global governance	1
Domain Name System	9	ICANN role	1
Privacy	8	Interconnection	1
Freedom of expression	7	Internet freedom	1
Human rights	5	Internet governing bodies	1
Multistakeholderism	5	Internet use	1
Openness	4	Juridical conflicts	1
Open standards	4	Malware	1
Diversity	3	Multiculturalism	1
Intellectual property rights	3	Net neutrality	1
Infrastructure	2	Technical standards	1
Internet Governance Forum	2	Routing management	1
IP addresses	2	Spam	1
Multilingualism	2	Youth	1
Public policy	2	Control mechanisms	1
Developing countries	2	IETF	1

ICANN = Internet Corporation for Assigned Names and Numbers; IETF = Internet Engineering Task Force.

shown in Table 5.1, we can certainly say that actors' participation contributes to enrich the meaning of IG, but this by itself does not say much about how this richness is translated into more articulated frames or how different themes and strands of reflection are connecting with one another to shape the thematic boundaries of the domain. What becomes more crucial, in this context, is the examination of how thematic inputs relate to one another to provide the bases for shaping collectively the IG discourse. We will try to investigate the enmeshment of different thematic inputs at two levels: on the one hand, at the individual level, to identify the different orientations converging in the construction of the discourse; on the other hand, at a more general level, to analyze how different inputs come together and shape the contents of the IG discourse.

1.1 Classifying Orientations toward Action in the Internet Governance Domain

When exploring how the thematic richness characterizing the contemporary IG discourse can be organized in terms of different "interpretative packages" (Gamson and Modigliani 1989), a first challenge consists in the identification of general dimensions around which the multiplicity of answers provided by our respondents can be organized. Adopting a bottom-up procedure that starts from the general IG definition provided by the WGIG (see Chapter 3), we propose to organize interviewees' inputs in relation to three main dimensions that jointly contribute to uncovering actors' understanding of IG: a *construction* dimension, which relates to *what* should be the object of governance processes; a *strategy* dimension, which relates to the *tactics* for improving the IG discourse; and, finally, a *development* dimension, which relates to the *ultimate goal* to be achieved through the deployment of the Internet system.

In general, answers provided for each dimension show the coexistence of different, to some extent opposing, perspectives. When looking at *what* is the object of IG (i.e., to the *construction* dimension), answers polarized around two positions: on the one hand, technological artifacts (e.g., Domain Name System [DNS] management); on the other, more socially oriented issues (e.g., "young people online"). As far as *tactics* are concerned (i.e., the *strategy* dimension), respondents emphasized two strategies: on the one hand, the affirmation of general principles that should be brought into and stated in the IG discourse the way they are (e.g., "freedom of expression," "respect of individual privacy"); on the other, a more process-oriented attitude emphasizing the promotion of dialogues wherein meanings are defined progressively and collectively (e.g., "finding common solutions to guarantee freedom of expression"). Finally, considering the ultimate goal of the Internet system deployment (i.e., the *development* dimension), two different scopes were identified: on the one hand, the reinforcement and the strengthening of the existing system (e.g., "improving security measures"); on the other, its growth to include previously excluded subjects, thus filling in multiple digital divides (e.g., "expanding Internet coverage to the South of the world").

In general, the answers provided by respondents can be considered indicators of underlying orientations that guide actors' participation in the construction of the IG discourse. If we refer the orientations to the three dimensions we just outlined, we can distinguish between more *static* or more *dynamic* orientations toward action. If we consider the *construction dimension*, when actors state that the main object of IG is technological, they communicate a preference for a more consolidated, traditional outlook on IG tied to the management of resources. Conversely, if in their answers they include broader socially oriented issues among IG thematic components, they dynamically open up to the enlargement of the agenda and to the management of various non-technical, and very likely controversial, issues. Looking at the *strategy dimension*, a preference for the inclusion in the IG discussion of already formulated principles conveys an idea of steadiness, as claims are perceived to be "out there," defined and crystallized in existing concepts that need to be incorporated into the discussion. On the other hand, answers explicitly suggesting the need for creating processes to collectively define claims, needs, and perspectives convey a sense of higher dynamism to the development of the IG discourse, which is seen as a work in progress. Finally, looking at the ultimate goals of IG, when respondents state that the final goal is to reinforce the existing system, they suggest a conceptualization that is somehow "limited" to the status quo of the Internet communication, whereas, when pointing toward more developmental aspects, they seem to hold a more dynamic, broad understanding that is connected to an idea of expansion, not just reinforcement, of the present status of the Internet system.[2]

Because IG is such a multifaceted domain wherein different and opposing elements can definitely coexist, individuals can have mainly static or very dynamic views, but, more often, they mix these two attitudes, for example, when acknowledging that both DNS and processes to guarantee freedom of expression are equally part of the IG discourse. Therefore, between the two opposite attitudes of steadiness and dynamism there are a multiplicity of "middle-ground positions" whose diffusion can give us an idea of how more traditional and more innovative conceptualizations of the IG discourse are actually meshing with each other. To understand the different proportions of steadiness and dynamism for each dimension, every answer provided by interviewees can be evaluated along the three abovementioned dimensions, and, for each of them, it can be assigned a value of +1 when communicating a dynamic orientation, a −1 when suggesting a more static attitude, or a 0 when not expressing any explicit preference. The sum of scores for each dimension provides the degree of steadiness and dynamism characterizing our respondents on each dimension (Table 5.2).

Negative values correspond to different levels of a predominantly static orientation, while positive values are associated with different levels of dynamic attitudes. As far as the *construction dimension* is concerned, it is interesting to notice that the majority of individuals tend to have a more dynamic orientation (48 percent) but that the prevailing single positions are the more moderate ones (−1 and +1). This suggests that there has been a contamination

Table 5.2.   Frequencies of the three orientation dimensions (*N* = *48*)

	Construction	Strategy	Development
−3	6.3%	16.7%	20.8%
−2	0.0%	4.2%	12.5%
−1	35.4%	27.1%	27.1%
0	10.3%	6.1%	16.6%
+1	29.2%	29.2%	16.7%
+2	4.2%	2.1%	2.1%
+3	14.6%	14.6%	4.2%
Total	100%	100%	100%

between the technological and more socially oriented discourses. However, if we look at the two extremes, dynamism (therefore, social issues) tends to prevail on purely technological matters. In the case of the *strategy* dimension, generally negative values are prevailing (48 percent); hence, the more static affirmation of general principles is generally preferred to the realization of processes. This trend is reinforced by extreme values where steadiness tends to prevail over dynamism. Still, also in this case the more diffused scores are +1 and −1, showing that preferences for general principles or for more process-oriented solutions tend to be not so neat. Finally, as far as the *development* dimension is concerned, the static orientation (meaning a focus on the implementation of the existing system) is prevailing over more developmental perspectives. Indeed, 20.8 percent of respondents focus totally on implementation of the existing network and, therefore, a predominantly static orientation, against the 4.2 percent of the predominantly dynamic category. Interestingly, in comparison with the others two dimensions, values of the neutral position (i.e., 0 score) in this case are the highest.

The overall predominance of more broad social concerns on the *construction dimension* should be interpreted as a consequence of the openness of the IGF environment, but yet, as shown by the widespread adoption of middle-ground positions, not as an indicator of a total reconceptualization of the discourse. Interestingly, the predominance of moderate positions along the *strategy dimension* indicates the coexistence of two trends that are only apparently in contrast with one another: the general prevalence of the static orientation might be connected to the necessity to affirm as clearly as possible actors' identities within the open conversational environment, but, given the dialogic feature of the space, defined symbolic statements are accompanied by an overall will to confront and, possibly collaborate with, other participants. Finally, more neatly defined results are obtained in relation to the *development dimension,* for which the empowerment of the present Internet system constitutes the main object of attention. Still, the rather high level of neutral positions in this case suggests that the other two dimensions are more relevant in structuring individual frames: instead of focusing on "where do we go from here?" it seems that individuals are trying to understand "where they are now" in terms of contents (construction dimension) and in terms of process (strategy dimension).

1.2 Semantic Networks on Internet Governance

As shown in Table 5.1 above, the joined participation of different actors in the construction of an IG discourse led to an enrichment of the overall discourse through the inclusion in the IG agenda of many thematic inputs. However, as mentioned, thematic richness per se does not say much about how the IG discourse is actually taking shape. To answer this question, rather than looking at the simple presence of an issue in the agenda, it is necessary to focus on how it relates to other themes and how it is integrated within the overall discursive dynamics that are taking place. In this sense, the evolution of discourses on IG can be better understood through the perspective of semantic networks. Figure 5.1 shows the overall network resulting from relations among the themes illustrated in Table 5.1: the presence of a tie indicates that at least one interviewee mentioned two issues together, and it can be read as a positive association between issues, while the absence of ties indicates a disconnection between themes.

The graph reproduces the enlarged IG agenda represented in Table 5.1, but it also shows clearly that the collective construction of the IG discourse is transcending the boundaries of thematic areas in a cross-fertilization of themes that is well represented by the numerous ties (i.e., thematic associations) that are joining together different nodes (thematic inputs). At first sight, though, this cross-fertilization seems even overwhelming, and no clear directions for discourse shaping seem to emerge from our network. Still, if we look at the thickness of the ties, which represents the number of individuals that have associated two themes in their answers, then discursive patterns are more easily uncovered. When the number of respondents associating two issues is higher than one, that is to say, when we consider only those thematic associations that are consolidating more remarkably within the IG discourse, the number of edges in the semantic network drops from 158 to 36, and most themes disconnect from the network, thus revealing the core of discourse formation (Figure 5.2).

In the first place, the meaningful decrease in the number of connected nodes in the network shows that, although all inputs legitimately enter the discussion, they do not strongly integrate within conversational patterns and remain simply occasional claims that could be raised thanks to opportunities provided by the open environment of the IGF. But after these "corollary themes" are disconnected, our network shows more clearly that an actual enmeshment between technical and socially oriented issues is actually taking place. Indeed, the right side of the graph groups more traditional, resources-related IG issues (e.g., t20ip "transition from IPv4 to IPv6," t8dns "domain name system management"), whereas more socially oriented issues are grouped on the left side: access, openness, diversity, open standards definition, and intellectual property rights, together with the recent and innovative issue of multistakeholderism (respectively, t1access, t28openness, t7diversity, t29os, t21ipr, t24msh). Interestingly, human rights (t11hr) enter the discussion through more focused arguments on freedom of expression. It should be noted that it is around ICRs, still central and at the core of the network, that this integration between traditional and recent topics is actually happening. In fact, if we consider exclusively stronger ties, the crucial role

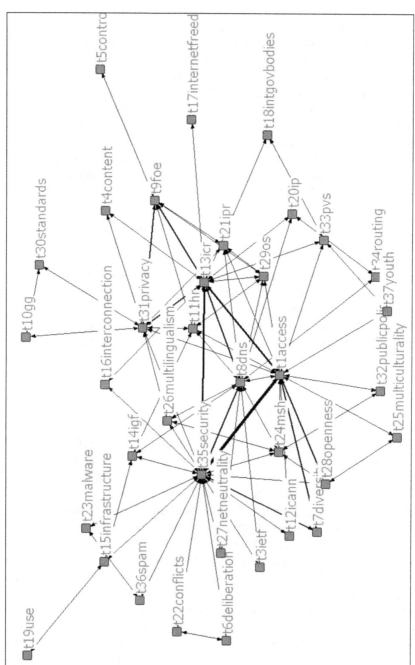

Figure 5.1.

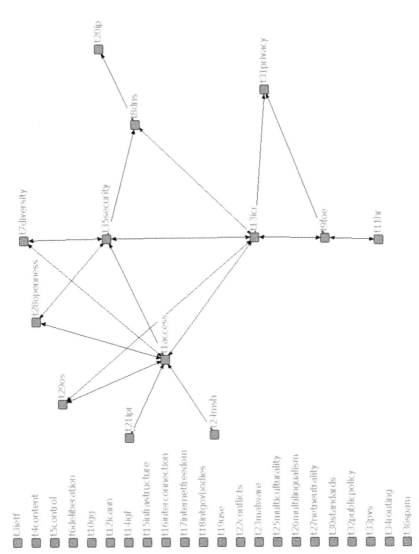

t3ietf
t4content
t5control
t6deliberation
t10gg
t12icann
t14oif
t15infrastructure
t16interconnection
t17internetfreedom
t18intgovbodies
t19use
t22conflicts
t23malware
t25multiculturality
t26multilingualism
t27netneutrality
t30standards
t32publicpolicy
t33ivs
t34routing
t36spam

Figure 5.2.

of ICRs is further confirmed, even though they are put in relation with other relevant issues. Indeed, the only conversational pattern involving at least five individuals is the triangle between access, security, and ICRs, which are not only the three most popular issues but also those that are more often discussed in connection with one another (indeed, their relevance is reflected in the structure of the IGF thematic schedule; see Chapter 3).

An even better understanding of agenda enlargement dynamics emerges if answers are grouped into subfields "which categorize topics of interest to particular participants" (Knoke et al., 1996:14). In this sense, six subfields can be identified starting from the answers provided: *security* (issues connected to security and privacy of Internet users); *Internet critical resources* (ICRs, infrastructural issues concerning management, development, and communication protocols); *access and use* (access to the Internet system, its expansion, and the inclusion of specific categories of users); *openness* (open and free communication within the Internet, interoperability, and development of standards); *human rights and freedoms online* (freedom of expression, freedom of speech, and intellectual property rights); and finally *governance*, probably the most heterogeneous subfield (political and legal arrangements, global governance, cooperation among stakeholders, implications of technology development for traditional political practices, and reform of IG mechanisms).[3] Overall, the subfield on security issues draws more attention than any other, followed by traditional discussion on Internet resources and by the whole range of issues connected to access (Table 5.3). Quite interestingly, subfields on thematic items more linked to broader social issues, such as openness of the Internet, human rights, and the actual reform of governance processes at large, appear less popular in comparison to other, more specific, concerns.

The analysis of relations between subfields confirms the pattern identified through the analysis of single thematic associations (see Figure 5.3): the "iron-triangle" of access, security, and ICRs is reproduced here as the very core of discussion. Yet, within this core, two privileged conversations are ongoing: the first, between the security and ICR subfields, suggests that the IG

Table 5.3.   Subfields of Internet and their relative importance identified from interview answers

Subfield	Size
Security (SECURITY)	28
Internet critical resources (ICR)	26
Access and use (ACCESS)	21
Openness (OPENNESS)	14
Human rights and freedoms (ARHF)	13
Governance (GOVERNANCE)	12

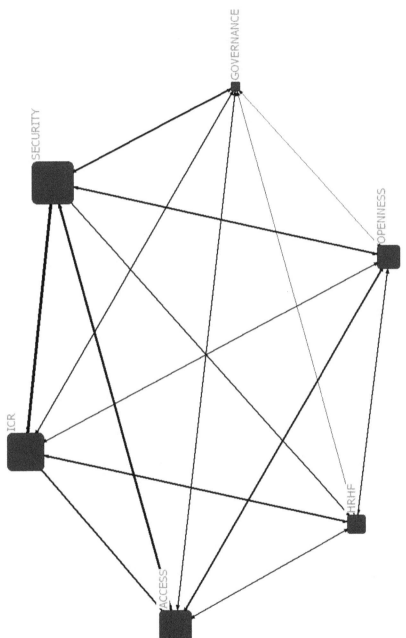

Figure 5.3.

discourse is orienting toward the (possible) implementation of the existing system; the second, between access and security, fosters reflections about threats posed by the very expansion of the system. The connection between the access and human rights subfields is weak in comparison to the others, and this suggests that access matters are more often discussed in connection with security concerns than in terms of human rights. Finally, it is worth mentioning that the governance subfield has very weak connections with the human rights one, as well as with the subfield pertaining to openness, and these elements, in turn, seem to suggest a tendency to avoid associating discourses concerning political procedures and processes to issues dealing with more general principles.

2. BIRD OF WHICH FEATHER? OFFLINE COLLABORATION NETWORKS ON INTERNET GOVERNANCE

The IGF process constitutes an unprecedented institutional occasion for networking. Indeed, in this context all formal constraints to participation are removed and the interest in the IG domain becomes the principal selection criterion for determining who is legitimately allowed to take part in the discussion. As a result of such an open environment, actors' behaviors and networking logics could result in a multiplicity of interaction scenarios, ranging from the total absence of discussion to a situation in which all actors involved speak to everyone else. Neither of these two extreme scenarios is correct. On what bases, then, is the collective construction of an IG discourse taking shape? In other words, who is talking to whom, and why? In a context characterized by a high level of heterogeneity of issues and actors involved, where there is a high risk of not recognizing other participants, either because they have different visions or simply because "they come from another group," we expect "homophily" mechanisms (McPherson, Smith-Lovin, and Cook 2001) to be guiding the partnership choices made by actors involved in order to reduce the uncertainty and complexity that characterize the overall interactional milieu and to enhance the effectiveness of their participation in the discussion. Nevertheless, if "homophily is the principle that a contact between similar people occurs at a higher rate than among dissimilar people" (McPherson, Smith-Lovin, and Cook 2001:416), it still has to be determined what makes individuals similar to one another in the IG domain given that status differences are abolished, at least in principle, and similarities and differences should be then referred to a more cognitive level.

Social networks and social network analysis can then be useful tools for clarifying the logics and patterns that are being followed when we look at the multiactor relationships that are built up in this domain. In the following sections, we will explore what kinds of social relations are being set up between actors of different natures in order to jointly contribute to the formation of an IG discourse. In particular, we will do so through the employment of the concepts and measures of *structural equivalence* and *centrality*, thus reinterpreting their traditional meaning in the context of mediated interaction (see section below entitled "A 'Blurred Duality'"). Indeed, mapping social relations at the transnational level is a rather challenging task, and there is a need to reconcile the theoretical potential of network tools and measures with the actual difficulty of having all relevant actors included in the examination. In this sense, without any presumptions for renewing social network analysis, here we will apply techniques, concepts, and analytic tools in a slightly revised version, so as to minimize the losses that derive from the hard task of gathering data on personal exchanges at the transnational level and to best exploit their explanatory potential.

2.1 Building a Transnational Network?

In order to map the relationships existing between individuals in the IG domain, the questionnaire contained a set of items that invited our interviewees to identify their partners.[4] When identifying a partner, respondents were also asked to specify the contents of the relationship with him or her.[5] As respondents could specify the contents of their relationships by choosing more than one type of relation for each partner, the overall strength of a tie between two individuals is provided by the sum of all contents specified and, therefore, can range from 1 to 6. On average, ties inside the network have a strength of 3, suggesting that relations in the domain tend to be composite and more than occasional.

Starting from our 49 respondents, a total of 138 partners were identified. The network involving these 138 "IG activists" is far from being complete, as information is missing for about 64 percent of nodes. Still, it is possible to examine this network from a transnational perspective (Figure 5.4) based on the geographical origin of our respondents. In fact, even in a very sparse network like this one, there is a center-periphery structure that is also North-centered (Katz and Anheier 2006). The majority of respondents, not very surprisingly, come from the European Union (EU) and the United States, and, with few exceptions, they all belong to the main component of the graph—together with Asian individuals—either thanks to direct mentioning or to indirect connections. The only Latin American representative is also part of the main component, actually as a result of the enhanced cooperation of Latin America with the rest of the world on the occasion of the second IGF meeting in Rio de Janeiro. African nodes, instead, form two separate components: not

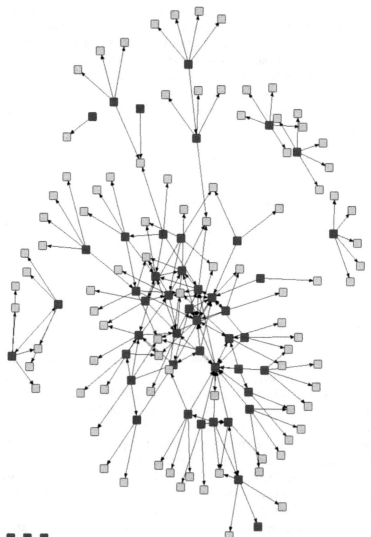

Figure 5.4.

only do African respondents locate their collaborations in a local (African) context that is not tied to the main component, but these two local interactional environments do not even overlap among themselves, as they are based in two different African states.[6]

2.2 A "Blurred Duality"

How can we exploit the relational data we gathered in spite of the lack of information for many of the partners identified by our respondents? Since considering only those nodes for which we possess information would give us only a partial overview of the situation, we can rely on Breiger's *duality principle* and *membership network* ideas (1974:181–193);[7] thus, starting from an affiliation matrix (N × M) in which individuals are lined up in rows (N) and their membership groups in columns (M), it is possible to obtain a matrix including only individuals where ties among them are defined by their presence in the same group.[8] The higher the number of groups to which individuals belong together, the stronger the tie between them. In our case, we can intend membership *metaphorically* and in relation to the action of partners' identification: whenever individuals identify the same partner, they "belong to the same group." On these bases, it is possible to build the network represented in Figure 5.5, which we call a "secondary or derived network," where individuals are tied to one another if they share at least one partner.[9] In this way, we end up having a network for which we do possess systematic information for all nodes, and, therefore, we can analyze multiactor partnership strategies starting from here.

However, there is one point that needs to be clarified. The network in Figure 5.5 derives from an affiliation network characterized by a peculiarity we could call a "blurred duality." Indeed, affiliation is traditionally referred to individual participation in the same group or event, but the two sets of nodes (individuals and events) are generally of a different nature.[10] The resulting affiliation networks are, therefore, non-dyadic (because relations join actors to events and vice versa) and dual (because of the complementarity of perspectives) (Faust 1997). However, in our specific case, events in the network joining together different respondents are intended metaphorically because they represent the common identification of the same partner. Therefore, the initial duality is not so neat, since there is no substantial difference between the elements in rows and columns (i.e., they are all individuals). Although blurred, this duality remains meaningful (rows provide us with information on the choices made by our interviewees—a proxy for integration in the emerging field—while columns provide information on the choices received by individuals identified in the field—a proxy for prestige). Our secondary

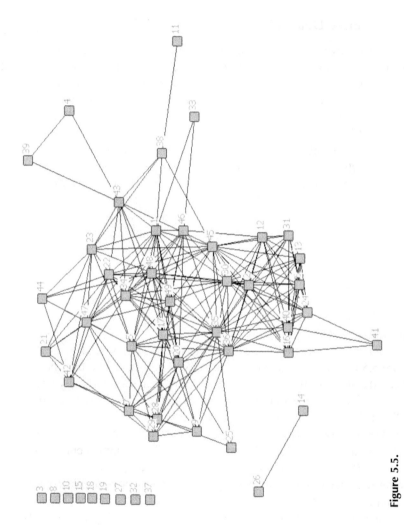

Figure 5.5.

network, then, brings with itself information always referred to individuals (their degree of activity and their prestige), and it is on this feature of blurriness that it provides a first step for analyzing the progressive construction of collaborative relations between different groups of actors.

With regard to our initial question, namely, what makes nodes similar to one another in a multiactor environment, our secondary network already provides us with some interesting insights. Nodes that are tied to one another can be considered similar because they share some of their partners. In contrast, isolated nodes can be thought of as different because, whether they have partners or not, they do not share them with anyone else. The underlying assumption here is that, in an open environment, collaborations are more likely to be established with actors that are recognized as legitimate partners. If two individuals do recognize someone else as a common partner, they surely have something in common, and communication between them could happen more easily.

A first perspective from which this network can be read, therefore, is that gained by exploring the reasons for isolation. In the first place, there are individuals who have agreed to answer the questionnaire without providing relational data (either because, acting as representatives of organizations, they could not identify single individuals as requested but only collective partners or because of privacy concerns). The second reason for isolation consists in localization of collaboration: individuals who tend to locate their partners in their neighborhood or daily interactional milieu become isolated in this secondary network. Conversely, those who identified their partners within the IGF process or, more in general, in the framework of international processes on GCG, belong to the main component. This is very much consistent with what emerged above by looking at the sparse network set up by all 138 IG activists who were identified through the interviews: there is a center-periphery structure in the dynamics of collaboration that are established in the IG domain. What is interesting to notice is that the more integrated discursive strands are those that develop transnationally within or in relation to international processes, whereas local discussions are basically disconnected.

However, the majority of nodes (38 out of 49) are part of the main component, and, therefore, it seems that a core of partnership is shaping. Still, the strength of these ties has to be critically addressed. If we raise the number of shared partners, the situation dramatically changes. When tie strength is set higher than zero, the network is composed of 412 ties (Figure 5.5) but, if we raise this threshold by only a unit, the number of ties dramatically falls to 58 and when ties strength is greater than 2 only 8 ties remain active. No actors have more than 3 partners in common. This dramatic decrease

Table 5.4. Blocks of structurally equivalent nodes in the derived network

Blocks	Nodes ID
1	{1, 30, 33, 34, 46, 48,49}
2	{5, 7, 9, 17, 20, 25, 28, 9, 35}
3	{21, 42, 44, 47}
4	{22, 23, 43,38}
5	{2, 6, 12, 13, 31, 36, 45}
6	{16, 24, 40, 41}
7	{3, 8, 10, 15, 18, 19, 27, 32, 37}
8	{4, 11,14, 26, 39}

seems to suggest that, if a core of partnership is in formation, it is still in its initial stage.

Going one step further, we can look for structural equivalence patterns (Lorraine and White 1971) or, in other words, at whether there are groups of individuals that are more similar among themselves on the basis of the relations they establish with others in the network. In this sense, we identify groups composed of more similar nodes where, once again, similarity depends on the partners shared. Within our networks, 8 main blocks can be identified (Table 5.4).[11] Looking more closely at the characteristics of individuals within each block, as well as at whether there are relations between blocks—which in this context can be interpreted as a higher degree of similarity between groups of structurally equivalent actors (Table 5.5 and Figure 5.6)—we are able to highlight some interesting patterns of partnership building.[12]

Overall, within our collaboration network there seem to be two main logics guiding the construction of (common) partnerships. The center-left side of the network in Figure 5.6 (blocks 1, 2, 3, 4) shows more clearly that

Table 5.5. Relations between blocks

Blocks	1	2	3	4	5	6	7	8
1	1	1	1	1	1	0	0	0
2	1	1	1	0	1	0	0	0
3	1	1	1	1	0	0	0	0
4	1	0	1	1	1	0	0	0
5	1	1	0	1	1	1	0	0
6	0	0	0	0	1	1	0	0
7	0	0	0	0	0	0	0	0
8	0	0	0	0	0	0	0	1

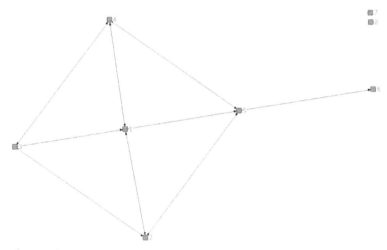

Figure 5.6.

there is a marked partnership-sharing mechanism and, therefore, a nucleus wherein conversational dynamics are deploying rather intensely. This part of the conversation between actors is structured upon long-term trust dynamics, built during the WSIS process but probably going even farther back in time. In this sense, collaboration within the IGF context revolves around privacy and freedom of expression concerns (adherents to the Privacy and Freedom of Expression [FoE] coalitions are located mainly here) and is grounded in previous collaborations started within WSIS caucuses (Privacy; Human Rights; and Copyrights, Trademarks, and Patents Caucuses). For this part of the network, then, multiactor collaboration within the IGF process looks more like the continuation of previous processes that have supported processes of long-term mutual recognition and fostered the enmeshment between different thematic focuses (see exchanges between blocks 1 and 2, the first concerned with privacy, the second with freedom of expression). As far as the right side of the network in Figure 5.6 is concerned (blocks 5 and 6), a more instru-mental logic of partnership building seems to be present. Indeed, the right side of the network is structured mainly around thematic commonalities on open standards and access to knowledge. Partnerships tend to be built within respective Dynamic Coalition (DC) environments and only rarely move beyond to reach the rest of the network. When this happens, it is the matter of access (mainly dealt with in block 5) rather than that of open standards (dealt with in block 6) that allows a channel for communication with the rest of the network.

2.3 From Leader to Catalyst: Centrality in a Partner-Sharing Environment

If we consider that similarities between actors are established within a conversational context (for the IGF is basically a multistakeholder space of dialogue), they can be considered also as elements facilitating communications among nodes and, ultimately, the construction of the IG discourse. In this sense, the concept of *prominence* provides a particularly useful entry point for examining how partner sharing translates into a concrete potential for contributing to the construction of the IG discourse. An actor is defined as prominent when she is "extensively involved in relationships with others" (Wasserman and Faust 1994:173). Within an interactional milieu like the one we are examining here, being extensively involved in exchanges means sharing partners with a larger number of other nodes in the network and, in being more similar to them, being part of more dense conversational dynamics. Moreover, as sharing partners is a reciprocal type of relation, in deciding how prominence should be measured we should prefer the concept of *centrality* to that of *prestige* (Knoke and Burt 1983), as the former recalls simply the involvement of an actor within exchanges while the latter is connected to the level of recognition one node receive from the rest of the nodes in the same network and, therefore, would require direct data.

However, given the indirect feature of ties in our secondary network, the very meaning of centrality needs to be slightly reassessed to be more consistent with the idea of similarity rather than to that of higher activity (Faust 1997). Thus, reassessing the meaning of prominence and, in particular of centrality, is far from being a mere exercise of rhetoric: rather, it is a crucial step, considering the strict tie existing between centrality and behavioral patterns (Coombs 1964). More particularly, because what we are examining here is a specific type of communication network, we lean on the conceptualization of centrality as tied to communication flows established between nodes (Freeman 2002 [1979], Friedkin 1991) and that elaborates on the communicative function played by each node given the actor's degree of involvement, her level of independence from the others, and her ability to control communication fluxes by acting as a gatekeeper. Since communication in our network is not direct (it could be interrupted by the withdrawal of the common partner), it is necessary to not only adapt the idea of centrality to the one of similarity but also to interpret it in terms of a *communicative potential* rather than of actual involvement in communication flows.

Table 5.6 provides a short overview of the traditional and the contingent meanings of some of the most popular centrality measurements interpreted in relation to communication processes within networks. Degree centrality

Table 5.6.   Traditional and contingent meanings of centrality measures performed on our secondary or derived network

	Traditional Meaning	Contingent Meaning
Degree: Freeman Degree	Activity or involvement in communication flows	Potential for direct communication among network members
Closeness: Geodesic Path Centrality	Distance of an actor from all others in the network	Size of mediation necessary for the actor to spread his or her message in the network
Betweenness: Flow Betweenness	Control over communication flows (measured on all possible paths)	Relevance of the interactional milieu brought by an actor

measured through the classical Freeman approach (i.e., the degree of involvement in communication flows) gives, more than other measures, a sense of a potential homophily mechanism underlying our network structure and, therefore, a more explicit indication of actors' communicative potential. In our network, degree centrality tells us how similar one actor is to others given the extent to which her friends are also someone else's friends. Thus, if the traditional measure conveys an idea of involvement in communication flows, in this case the index stands for the potential for communicating with others in the group: *the higher the similarity level, the higher the potential for communication within the construction of the IG discourse.* Closeness centrality focuses on the distance between actors in the network: the higher the distance the higher the dependence of actors from others in communication flows.[13] In this case, the highest value means that more intermediaries are needed to connect the actor to everyone else in the network. In other words, actors with the highest distance from others are those who need more mediation to spread their messages. For us, *the higher this index, the less the efficiency and rapidity with which actors are able to exert influence in the construction of the discourse.* Finally, moving to betweenness, it is the original idea of mediation or control over communication flows that must be reframed here. Indeed, as every tie in the network we are analyzing is indirect, it already carries with itself a sense of intermediation (which is exerted by common partners). Nevertheless, betweenness measures can still tell us something about the importance of the interactional milieu created by one actor in the network. In our specific case, *the higher the betweenness value, the higher the relevance of the partners' set brought by the actor in the network.* In other words, if that actor were in a different position or were excluded from the network at some point, communication flows would be less easy, although not absolutely impossible. When we look at betweenness measured on all possible paths between nodes (flow betweenness), we amplify this perspective by discovering the role played by all actors in the broader structure and not only in the immediate surroundings.

If centrality measures change their meaning, then also roles connected to central positions in the networks need to be reframed. When relations are indirect, as in this case, actors' power can be linked to their capacity to speed up, foster, and facilitate interaction between different participants, thus contributing substantially to the consolidation of discursive multiactor relationships. In doing so, these actors works as *catalysts* and distinguish themselves from others in the same interactional milieu as they join together leadership, brokerage, and entrepreneurial features and operate in a context in which ties among actors are not direct but mediated by third individuals.[14] Catalysts do not directly exploit their position: they facilitate communication flows between others, as they are in a more favorable position to collaborate with others in the network. Much like leaders, catalysts are able to pool resources and promote action inside the network within which they cover key positions (Diani 1995, 2003; Schou 1997); like brokers, they do foster communication and interaction between different groups (Fernandez and Gould 1989, 1994) and invest time and resources in mediating between other network members while foreseeing some kind of personal profit resulting from mediation (Boissevain 1974); also, they create opportunities by actively engaging in uncertainty reduction (Christopoulos 2006) and adopting entrepreneurial behaviors. Overall, catalysts stimulate the production of the IG discourse by setting up a favorable environment thanks to their position (Knoke 1990; Knoke et al. 1996; Laumann and Knoke 1987). Starting from the reviewed meaning of centralities we gave above, catalysts are actors in the network that are more similar to others because they share their partners with a high number of other members; they need less intermediation to spread their messages and, therefore, are more effective and quick in achieving their task of fertilizing the field for an actual dialogue; and they facilitate communication in the network because, if they left, dialogue could still be possible (their absence does not mean the cancellation of all interaction in the network), but it would be harder, slower, and probably less effective.

Who is playing the catalyst function in the construction of the IG discourse? Table 5.7 summarizes centrality scores conducted on our network with specific regard to nodes' potential for direct communication (i.e., the Freeman degree), the amount of intermediation needed (i.e., closeness), and the relevance for the interactional milieu created (i.e., flow betweenness).[15] Even at a first glance, our network does not seem significantly different from the "latent network" profile derived from literature (Diani 1995; Melucci 1984): horizontality given by the high number of connections existing between

Table 5.7. Centrality scores measured on main component nodes (N = 38)

Actor	Degree	Closeness	Flow Betweenness
1	16.00	59.00	5.78
2	13.00	65.00	1.71
4	2.00	100.00	2.63
5	15.00	63.00	3.36
6	11.00	71.00	1.71
7	10.00	71.00	2.31
9	7.00	74.00	1.04
11	1.00	111.00	0.00
12	12.00	68.00	1.59
13	11.00	71.00	1.71
16	12.00	70.00	2.87
17	18.00	58.00	4.66
20	14.00	64.00	2.52
21	6.00	82.00	1.69
22	13.00	63.00	2.88
23	11.00	65.00	4.41
24	12.00	70.00	4.44
25	3.00	81.00	0.43
28	8.00	73.00	2.76
29	8.00	73.00	1.86
30	18.00	59.00	2.14
31	11.00	71.00	1.47
33	2.00	93.00	0.22
34	14.00	62.00	2.57
35	24.00	53.00	6.13
36	19.00	59.00	3.75
38	5.00	75.00	6.16
39	2.00	100.00	2.63
40	12.00	70.00	3.03
41	3.00	104.00	0.57
42	8.00	79.00	2.78
43	10.00	65.00	12.41
44	6.00	82.00	1.52
45	20.00	54.00	4.24
46	12.00	66.00	3.66
47	12.00	73.00	3.56
48	17.00	59.00	2.48
49	12.00	66.00	1.35
MEAN	10.79	72.16	2.92

nodes and not much inequality distribution inside the network. Nevertheless, individual differences still persist and deserve a closer examination.

Actors 1, 5, 17, 23, 24, 35, 36, 40, 45, and 46 are catalysts. Their high degree indicates they have the highest potential for establishing direct communication with other network members.[16] They are also those who spread more independently their messages in the network and tend to provide the most crucial interactional milieu. In other words, in being characterized by a greater communicational independence, catalysts are also those who tend to facilitate communication in the rest of the network. All these actors except actor 5 belong to the public interest sector, were actively involved in the WSIS process, and are predominantly interested in privacy and human rights (as testified by their DC membership), but some of them are interested in more technical issues such as open standards and the matter of access. At the opposite end, some other actors in our network (4, 11, 21, 25, 33, 41, 42, 44) participate in the positive exchanges in the network and contribute to the creation of the overall dialogue yet without determining it. It is interesting to notice that all these nodes in our network except 21, 41, and 44 do represent national or international institutions.

Actors 38 and 43 show relevant values in flow betweenness. If this measure stands for the propensity to set up a favorable communicative environment (given the overall network structure), it is interesting to see that these two nodes represent the International Telecommunication Union (ITU) and the Organization for Economic Co-operation and Development (OECD). Both actors' centrality patterns seem to suggest a potential for enhanced collaboration by institutional actors. Actor 49 also represents OECD, but its betweenness is not as high, whereas its degree centrality is higher than the mean: together with the propensity for setting up a collaborative environment, then, some hints of a potential direct multiactor communication are coming also from institutions.[17]

2.4 Explaining Centrality Patterns

What could help explain the centrality values of different actors and, therefore, their communication potential? To answer this question, we can focus our analysis on the Freeman degree and flow betweenness[18] to illustrate two different yet interrelated mechanisms that are grounding discursive activities: on the one hand, the potential for immediate collaboration (given by similarity levels) and, on the other, the potential for establishing a favorable interactional milieu.

In exploring the relevance of individual characteristics for increasing or lowering the potential for communication within our network, some elements seem of particular interest. In the first place, although much emphasis was put on the cancellation of status issues within the IGF, it is reasonable to think that the membership sector can have a role in fostering or hampering communication flows between individuals. Also, we can hypothesize that a longer experience with mobilization in this domain could affect one's potential to communicate to other, because, if it holds true that participants to the Forum are considered for what they say more than for who they are, then a more consolidated experience can provide individuals with higher authority and, hence, with a higher communication potential. Moreover, given the existing legacy between the IGF and previous international processes, an integral part of individuals' experience in the IG domain is provided by their participation in the WSIS because it provided the first stage for the development of a multiactor conversation on IG. Finally, we can think that the very orientations individuals hold in relation to the construction of the IG discourse (see early section in this chapter entitled "Classifying Orientations toward Action in the Internet Governance Domain") can influence their ability to speak to larger groups: more dynamic orientations could translate into a higher propensity to collaborate with others and, therefore, would increase the probability of having partners in common.

These considerations were tested on respondents' values of degree centrality and flow betweenness: first, we wanted to see the relevance that each of the abovementioned factors have on actors' centrality and, second, we looked for possible differences between the two communication potentials entailed by the two different measures. As shown in Table 5.8, as far as degree centrality is concerned, it is interesting to notice that membership sector by itself has no significant effect but that, when including participation in at least one of the WSIS meetings, to be a non-institutional actor favors higher scores. In other words, non-institutional actors participating in the WSIS are more integrated into our network and, hence, hold a higher communicative potential. This finding confirms that the WSIS actually marked a turning point in fostering a larger mobilization and involvement of public interest initiatives within the discussion and that civil society actors might have exploited it to a larger extent than institutions. Moreover, it is interesting to notice that the amount of experience in the field has no influence on the potential for direct communication, whereas the principal effect seems to be generated by orientations toward the IG discourse construction. Indeed, when we add indexes connected to individual frames, the model improves substantially. In particular, individuals who concentrate to a larger extent on IG social issues and those who are pushed more toward attention to the construction of processes, rather than on the affirmation of principles, are more central and, therefore, more integrated in the current partnership structuring dynamics. In

Table 5.8. Regression coefficients on degree centrality and flow betweenness

	Dependent Variable					
	Degree			Flow Betweenness		
	Model 1	Model 2	Model 3	Model 1	Model 2	Model3
Basic characteristics						
Membership sector	0.244	0.324**	0.236*	−0.239	−0.228	−0.288*
Experience in the field	0.202	−0.017	−0.006	0.406**	0.378**	0.371**
Participation in the WSIS	—	0.431**	0.379**	—	0.056	0.018
Orientations						
Construction scores	—	—	0.325**	—	—	0.186
Strategy scores	—	—	0.462***	—	—	0.441***
Constant (S.E.)	6.561*** (2.275)	5.136** (2.202)	6.115*** (1.866)	24.986** (11.247)	24.014** (11.806)	29.096*** (10.709)
R^2 Adj.	0.060	0.180	0.426	0.156	0.134	0.399

Note: WSIS = World Summit on the Information Society; * = $p < 0.1$; ** = $p < 0.05$; *** = $p < 0.01$; — = not included.

this sense, it seems that socially oriented issues provide a more fertile ground for dialogue building, probably by virtue of the wide range of positions that can be taken over them but, most of all, because these are issues that have been dealt with in the IG framework for a more limited time span and, therefore, need further confrontation and clarification. Secondly, the relevance of individual orientations toward process building, rather than toward closure on fixed principles, fosters the possibility of creating a collective IG discourse. This is not to say that principles do not matter any longer but, rather, that the ability to confront one another is a more relevant condition for the realization of a shared discourse.

As far as flow betweenness is concerned, we notice that the amount of experience is a constantly significant variable: the longer the experience, the more crucial the presence in the network. This finding suggests that, if the focus is on how a favorable communication environment can be created, then actors' status (i.e., who they represent) is less relevant than the possession of useful knowledge. Moreover, it is interesting to notice that in this case participation in the WSIS process is not a significant variable.

This finding suggests that, although the legacy between the WSIS and the IGF remains important for "recognizing" possible partners, participation in previous international debates is not a necessary prerequisite for being a relevant part of IG discourse at the present stage. A consistent finding is that a preference for socially oriented issues is not significant: what is relevant for setting up a favorable communication environment is "being there today," with an attitude toward process building, and hence, toward collaboration. In this sense, while commonalities on IG contents might make it easier to recognize others as possible interlocutors in the process, without an explicit attitude toward collaboration it is harder to actually build multiactor collaboration regardless of the contents brought to the table. But our findings suggest another interesting element. When the strategy dimension is added in the third model, the membership sector variable becomes significant but with a negative sign: when orientation is favorable to process building, being an institutional actor means bringing into the network a more relevant interactional milieu. This, in turn, is consistent with the fact that the Forum, regardless of how innovative it can be, remains an institutional gathering and responds, among other needs, to the necessity of finding widely agreed-upon ways of integrating institutions in the IG discourse. In general, then, if civil society is crucial for providing knowledge, institutions remain an important part of the game and need to be fully involved in the construction of the discourse.

3. A REFORM, NOT A REVOLUTION

In this final section we try to summarize and pull together the results we derived from the analysis of social and semantic networks to obtain a snapshot of how the IG discourse is being constructed offline by the joint collaboration of institutional and non-institutional actors. Both levels of analysis showed that the current offline situation of the discussion is carried on in the context of the general heterogeneity that is characterizing thematic understandings and is fostering the creation of collaborative synergies.

Looking at frames and themes, we outlined in the first place the main individual orientations guiding the construction of an IG discourse in relation to the objects of the governance process (*construction dimension*), the tactics that should be adopted (*strategy dimension*), and the final scope of the development of an IG discourse (*development dimension*). Secondly, we moved to study how thematic inputs brought by different participants converge into networks of thematic association to shape the directions of the agenda enlargement processes in the IG domain. Although carried on at two different levels (i.e., the individual and the discursive), our analyses focused on the semantics of IG show an unquestionable enmeshment of thematic strands in which more technical matters are mixing up with more socially oriented concerns. At the individual level, the diffusion of orientations that combine the static affirmation of crystallized principles with a more dynamic and process-oriented discussion shows that the discourse on IG is *in fieri* and that there is a relatively growing tendency to accept a redefinition of the thematic boundaries of the domain thanks to open collaboration with other actors. Moreover, at the discourse level, semantic network analyses show that thematic inputs brought by actors participating in the discussion are coming together in a cross-fertilization of areas that generally enriches the IG agenda well beyond technical matters. Thus, the development of a more socially oriented side of the IG discourse seems to overcome even reflections on technical matters. At the individual level, this predominance is shown by the larger diffusion of more dynamically oriented attitudes on IG contents; at the level of discourse, instead, the study of semantic networks shows the actual integration within conversational dynamics of a large number of socially oriented issues through the establishment of stronger thematic associations.

However, at a closer look, the individual and the discursive facets of our analysis invite us to weight carefully the results we obtained. Indeed, at the individual level, despite an overall predominance of social issues, participants endorse more often moderate positions that integrate the two main perspectives on IG (i.e., the social and the technical). At the level of the overall discourse, instead, the integration of different perspectives is actually happening around more conventional concerns, namely, the management of Internet critical resources. Moreover, looking at thematic subfields in the broader domain, semantic network analysis showed that, if socially oriented issues are gaining importance, this happens mainly under the banner of security and privacy issues while other generally socially oriented subfields are less strongly integrated into conversational patterns. More particularly, besides Internet critical resources, security issues and the broad matter of access provide the two main areas around which the discussion is developing. The discussion on other issues revolves around this thematic triangle, but the integration within it of more broader concerns, such as human rights and human freedoms, is weaker. Not only social issues enter the agenda under the banner of security and privacy or in broader terms, such as "guaranteeing access." Security and ICRs are most commonly discussed together, in search of a balance between the management of technical elements and the protection of existing users. Thus, access is discussed more often in relation to security concerns than in connection with human rights matters, and this, again, is done in search of a balance that entails the protection of existing Internet users, rather than with the aim of filling in multiple divides. This last element is also confirmed by the study of individual orientations toward the construction of the IG discourse: the implementation of the existing Internet system, in general, prevails over concerns over its expansions. In this context, beside more technical and traditional matters we find that part of "social" discourse that could no longer be avoided: how to guarantee privacy and security, how to ensure the broadest access possible without risking the collapse of the system or its overall ineffectiveness. The rest of a socially oriented discourse remains, at present, secondary. Overall, then, more than toward an agenda revolution, multiactor discursive dynamics seem to be oriented toward the search for a balance between three thematic cornerstones and related needs: guaranteeing security, managing critical resources, and providing broader access to the Net.

In the second part of this chapter we exploited the potential of social networks analysis techniques with the overall aim of exploring what kinds of social dynamics are sustaining multiactor dynamics for the construction of the discourse in the IG domain and for gaining a better insight into the roles of different actors in the conversation. Overall, efforts to reduce procedural uncertainty through conversational practices, thus taking advantage of actors' commonalities, were explored starting from the concept of homophily (McPherson, Smith-Lovin, and Cook 2001) or, in other words, those mechanisms by which actors tend to build ties to those who are perceived as more similar. We argued that, in a complex context such as the one we are examining here, where everyone is free to participate on an equal footing, it still remained to be assessed what is perceived as constituting an element of similarity between actors. To answer this broad question, we built a derived matrix in which respondents to our survey were tied to one another when they shared the same partners. Even though we lost information on the direct choices made by our interviewees, we could explore the construction of multiactor relations starting from a network of similarity where existent ties were interpreted in terms of similitude, given the fact that sharing the same partners entails having something in common, but also in terms of communicative potential because collaborating with the same partners can be considered an element that facilitates the collective construction of a discourse over policy matters.

In the first place, we examined isolated nodes to understand why they differ from others. In addition to a lack of information on collaborators, isolation is mainly due to partnership localization. In other words, when partners are identified within daily contexts and not in broader international spaces, actors in our networks tend to be more peripheral, if not isolated. As far as those who share their partners are concerned, the analysis of our network shows that there are two main strategies guiding the formation of partnership-sharing blocks and, therefore, the collective shaping of the IG discourse. On the one hand, starting from the WSIS experience, there is a long-term solidarity that has built up and that finds in the IG process a further occasion to consolidate. On the other hand, there is a thematic, instrumental strategy of partnership building that is based on commonalities of interest and for which the dynamics currently developing in the IG domain constitute the basis for the construction of new alliances sustained by thematic proximity. The first strategy is reinforcing the construction of conversations mainly on privacy and human rights, while the second supports the articulation of themes such as openness of standards and of technologies.

To a further level, collaboration in the network was read from the point of view of positions occupied by single nodes and interpreted in terms of the communicative potential held by actors in the construction of the IG discourse. The *catalyst* figure was introduced to depict the role played by those actors who are facilitating the construction of the discourse and who, in doing so, join together qualities that are associated with leaders (resource pooling), brokers (exploitation of strategic positions within the network structure), and entrepreneurs (adoption of behaviors oriented to uncertainty reduction). In general, the catalyst function is played by non-institutional actors who were previously active in the WSIS context and currently focused on social matters (such as privacy or human rights at large).

We finally moved to examine more closely what elements foster or hamper the communication potential of individuals, and we did so by exploring possibilities for establishing direct dialogue relations (Freeman degree) and the importance of the interactional milieu brought by actors (flow betweenness). In this sense, non-institutional actors showed a higher potential for communicating with others in the network. Also, a focus on the most uncertain part of the IG domain (i.e., socially oriented issues) tends to foster levels of integration within the discourse, and so does an individual orientation toward process building. Nevertheless, when it comes to the evaluation of elements that are contributing to the creation of a relevant interactional milieu, longer experience and an orientation toward process building are more important than actors' status (i.e., who they represent, whether institutions or non-institutional entities). And yet, among those who adopt a process-oriented attitude, institutional actors tend to be more crucial than others in the construction of an IG discourse, and this is particularly true in the OECD and the ITU cases. More generally, in evaluating the influence that frames and orientations have in fostering communication, our analyses showed that, while commonalities on IG contents might make it easier to "recognize" others as possible interlocutors in the process, without a positive attitude toward collaboration it is harder to actually build multiactor collaboration regardless of the content brought to the table. In this context, civil society seems to foster the circulation of knowledge, whereas much of the success in collectively constructing an IG discourse seems to be connected with an institutional collaborative attitude.

If we connect considerations on themes and on processes, we notice something interesting. As outlined above, two different strategies seem to be at work in structuring partnership relations in the field: on the one hand, long-term solidarity and, on the other, instrumental solidarity based on thematic commonalities. It is not by chance that security matters are discussed within a core of the interactional structure characterized by long-term solidarity and, at the same time, are the most populated thematic subfield in the current IG landscape. Thematic popularity and long-term solidarity go hand in hand and determine the consolidation of this thematic area in the IG agenda: it is the popularity of this topic, together with the fact that it actually touches the interests of all stakeholders, that fostered (perhaps forced) an early and sustained convergence on it from the WSIS on. On the other side, less popular or less crosscutting themes find in the current IGF context a first space where convergence can be realized, and, in this sense, coalitions are being formed on the basis of thematic proximity with the overall aim of securing their consolidation within the IG agenda. More systematic and longitudinal studies would be necessary to assess whether long-term and instrumental solidarity constitute two different stages in the consolidation of themes and alliances in the IG domain. It is reasonable to think that the latter correspond to an early stage of subfield consolidation, a stage in which actors previously disjoined start to mutually recognize each other and to evaluate the range of possible more solid alliances. The reiteration of interaction patterns based on thematic proximity, then, might lead to the formation of long-term coalitions joining together governmental and non-governmental actors on shared bases built on interest commonalities and preparing the ground for actual political collaboration.

However, we have noticed also that, despite the premises of the multistakeholder approach, not all actors' contributions to the construction of the IG discourse are actually equal. The analysis of offline collaboration networks confirmed that non-institutional actors do play a major role in conversational dynamics and seem to be those exploiting to a larger extent the possibility of participating in the debate. Messages sent by non-institutional actors are those that are less in need of intermediation to travel along network ties. If, on the one hand, the immediacy of their messages contributes to an unprecedented variety in the range of issues discussed, on the other, the multiplicity of competencies, perspectives, and expertise they bring into the picture is certainly lowering IG uncertainty, as the progressive consolidation of the thematic triangle ICRs-security-access at the core of the discussion shows. The current dynamics of IG discourse construction are definitely showing the real possibility of integrating institutional and non-institutional actors and their perspectives within the same collaboration space, both enriching the overall debate and contributing to the establishment of common priorities.

Once again, this transformation is less revolutionary than it seems. Indeed, it is far from constituting a total ousting of traditional political actors from the scene. When it comes to evaluating how crucial is the interactional milieu brought by each actor in the network, that is to say, whether his or her absence would jeopardize a successful end to discursive dynamics, the governmental and intergovernmental actors who entered the discussion with a process-building attitude are more relevant than the non-institutional ones. If non-governmental actors are fundamental to the construction of an IG discourse because they easily communicate competencies and expertise to the rest of the network, institutional actors are also crucial in setting up conditions for this transmission to actually happen. Non-institutional actors contribute to the reduction of overall uncertainty, whereas institutional actors do provide a favorable environment for them to do so.

And yet, what is actually providing fertile ground to turn this mutual dependence into a reciprocal gain is that communications and collaborations seem to be built more on the basis of orientations and frames than on status-based considerations. The relevance of inbred attributes, such as membership sector, is more limited in comparison to attitudes that actors have toward the collective construction of the IG discourse. While reaching out to broader socially oriented issues increases the possibility of constructing inclusive and participatory dynamics, actors who adopt a process-building attitude are characterized by a higher potential for communication both in terms of possibilities to establish direct contacts and in terms of contributing to the creation of favorable interactional environments. In general, then, it is the progressive construction and acceptance of new logics of mutual recognition, based on reconceiving similarity and proximity in connection with understanding and political will rather than on baseline qualities, that influence the possibility to successfully collaborate in the IG domain. Formal openness, without an explicit will to grasp the opportunities provided, overall does not reform political arrangements.

NOTES

1. One interviewee did not actually provide any answer to questions in the survey concerning themes to be associated with the label of IG and, therefore, was excluded by analysis.

2. Steadiness should not be erroneously interpreted as a more negative attitude, nor should dynamism be understood as a more positive and desirable one: they represent different ways of *framing* participation in the construction of the IG discourse. Steadiness connects, mainly, with preferences for elements that are consolidated, acknowledged, and institutionalized and, therefore, indicates that individuals are less keen to further put under scrutiny a certain element. Conversely, dynamism connects with explicit preferences for fluid conceptualizations of themes and processes and, in this sense, simply suggests that individuals are available to further negotiate meanings, procedures, and goals.

3. Labels attached to every subfield are by no means exhaustively summarizing their contents but rather serve illustrative purposes: every theme in Table 5.1 has been assigned to one and just one subfield that has been labeled perfunctorily in order to summarize the contents included within it. Moreover, subfields differ in the number of categories they include. Each subfield corresponds to a dichotomous variable, and its size is given by a count of positive answers to the area it is identifying.

4. The questionnaire allowed for the identification of up to five partners. The average number of partners identified was four. Only four interviewees did not provide any partner's name.

5. The survey looked at five kinds of relation that can possibly be interesting for this field: (1) belonging to the same Dynamic Coalitions (DCs: thematic groups developed within the IGF context but possibly working also outside it); (2) sharing or exchanging information; (3) sharing or exchanging resources; (4) supporting initiatives; (5) realizing joined activities. Moreover, a sixth open possibility was provided to enrich the relational contents examined (mainly with formal employment or research relationships or membership in the Multistakeholder Advisory Group, the coordinating body of the IGF).

6. If this particular feature of our network could recall the North-centered feature already highlighted in the empirical investigation of transnational networks (Katz and Anheier 2006), it must be said that the administration of a questionnaire in English has further penalized relational data gathering by eliminating from the network French-speaking African participants.

7. This duality of individuals and groups goes back to Simmel (1955 [1908]), for whom individuals are differentiated from one another depending on the specific patterns of adhesions to groups characterizing them, and, at the same time, groups

can be differentiated from one another on the basis of the unique composition of the individuals participating in them.

8. The operation entails the multiplication of a matrix A (N×M) for its transpose A^{-1} (M×N), where original rows and columns are inverted. Of course, the same procedure can be applied to groups: it will be enough to transpose with regards to groups. The final matrix will have groups in both rows and columns, and ties will exist among them if they are participated in by at least one common individual.

9. There are three types of information that we lose in the operation through which we derive our secondary network. In the first place, direction of ties is absent, as the tie represents the presence of a common partner. Secondly, it is not possible to know directly from this network who the common partners are, and, in this sense, information about popular actors in the field is lost. Finally, it is no longer possible to determine whether two interviewees identified each other as partners.

10. See, for example, interlocking directorates phenomena (Bunting and Barbour 2002 [1971] and Mariolis 1975); the "duality of persons and groups" (Breiger 1974); and patterns of women's participation in different social events (Davis et al. 1959).

11. Blocks were identified through the CONCOR algorithm (Schwartz 1977). For the 8-block solution illustrated here the maximum depth of split was 3, diagonal values were ignored, and the resulted R value was 43.3, indicating a rather high goodness of the modeling solution obtained.

12. Relations between the 8 blocks were identified on the basis of a cutting-edge value given by the density of network in Figure 5.5 ($\Delta = 0.20$). Where the original exchange matrix showed a value higher than Δ, then a 1 was reported in the image matrix represented in Table 5.3. It must be remembered that the matrix represented in Figure 5.6 is a valued graph, that is, a collection of nodes and lines or arcs in which arcs carry a value that, in this specific case, is given by the number of common partners. The meaning and formulas for density change pass from a graph to a valued graph. "For a valued graph/digraph, the density is $\Delta = \Sigma \, v_k/g \, (g-1)$ where the sum is taken over all k. This measures the average strength of the lines/arc in the valued graph/digraph" [v_k is the value of ties in the graph, g the number of nodes] (Wasserman and Faust 1994:143).

13. As pointed out by Hannemann and Riddle (2005: ch.10), closeness can be operationalized in terms of farness, that is, in terms of distance between nodes in the network. There are several ways of calculating distance between nodes: in this case, we opted for a measure of farness calculated as the sum of all geodesics between nodes or, in other words, "the sum of the lengths of the shortest paths from ego (or to ego) from all other nodes" (Hanneman and Riddle 2005: ch 10). In this sense, the higher the value, the higher the distance from a node to all others.

14. Third parties mediating the interaction could nevertheless prevent direct relationships between two nodes. This is mainly the reason why the idea of power cannot be considered following traditional relational views.

15. Centralities were calculated only on the main component of the graph (38 out of 49 nodes). Flow betweenness was preferred to traditional betweenness as it is

calculated on all possible paths between two nodes, thus proving a better index for the relevance of the interactional milieu created by nodes in the whole discursive dynamic.

16. In this regard, it is important to notice that often these actors also do recognize one another directly, although this information is not directly available in the network we are analyzing (see note 11).

17. This is due to the fact that interviews have been done in two different moments in time, with actor 49 being interviewed during OECD efforts to set up a "Reference Group on Internet Governance" with experts coming from civil society.

18. All centrality measures are positively and highly correlated with one another. However, the correlation of the flow betweenness value with other measures was of lower intensity, thus suggesting the interrelation of corresponding social mechanisms, not their total overlap.

Conclusions

The voyage of discovery is not in seeking new landscapes but in having new eyes.

Marcel Proust

In concluding this work, there are three levels at which final considerations should be drawn. Adopting an inverse order than the one we followed in the book, the first level pertains to the specific case study on which this work has focused, that of Internet governance (IG). In this study we considered the IG domain as a part of the broader Global Communication Governance (GCG) field, and we engaged in a research effort aimed at deepening our understanding of what we believe constitutes a crucial but still overlooked domain in the contemporary global political landscape. We started our investigation from an unconventional point of departure: the intertwinement between IG contents and processes in the construction of a new, collective, and multiactor IG discourse in two different discursive spaces—the online and the offline. Leaning on an extended conceptualization of discourse (understood as a metaphor for social interaction between political actors; see Donati 1992), we explored through networks the discursive dynamics developing online and offline by looking at both the semantic and the social sides of the IG discourse. It is time now to read jointly the results obtained from the study of online and offline communication governance networks in the IG domain to derive an overall overview of this area.

The second level at which we should draw conclusions is the broader context of GCG and global politics reform. Because we consider IG one of the most crucial domains in the GCG field, we should try now to answer our initial and original question, namely, how the GCG field is being structured

151

and what the current governance arrangement in this area can suggest to us about the progressive reform of supranational politics. Finally, we should elaborate more generally on the appropriateness, potential, and limits of the approach we adopted to meet our research purposes. Indeed, we endorsed a network perspective and employed network analysis techniques to gain new insights on the construction of an IG discourse, and, in this way, we could actually account for the overlooked interplay between contents and processes that characterizes both this specific domain and the broader GCG field. However, in using networks to learn more about the object of our research, we also learned something about networks themselves and, more precisely, about the challenges that the adoption of a network approach to the study of supranational politics entails. It is by reflecting on these challenges and on the potential of a network approach to the study of supranational politics that we would like to conclude this work.

1. INTERNET GOVERNANCE: THE CASE

While at the end of Chapters 4 and 5 we summarized the results obtained from the study of online and offline semantic and social networks, now we join them together to derive a general understanding of the status of IG contents and processes.

As far as contents are concerned, this work contributed to systematically showing that IG is characterized by conceptual complexity and dynamism. Online and offline networks are coherently stressing the fact that the IG agenda no longer includes purely technical matters but actually develops around a rainbow of themes that connect together in a semantic maze. This finding might appear obvious to those who are already involved in the discussion, but it is something that cannot be overlooked. From a pure technical topic, IG has come to mean a plurality of intertwined themes and elements, and, in this sense, it is progressively being constructed as a social problem (Hilgartner and Bosk 1988). At present, the IG discussion comes to subsume a more broad discussion on the strict relationship between Information and Communication Technologies, or ICTs (which find on the Internet their epidermic infrastructure) and society at large. The thematic extension reached by the umbrella term of *Internet governance* is unprecedented, and, what is even more interesting is that such a richness and complexity of the discussion have been achieved in a very limited time span. On the one hand, this is a direct consequence of the fast Internet evolution. On the other, it points to the necessity of designing and adopting a flexible and efficient governance arrangement that can host a constantly enlarging discussion.

If we look more carefully into our online and offline semantic networks to examine how this agenda enrichment is actually happening, we notice that the construction of the IG discourse entails, precisely, *an enlargement and not a revolution.* Technical elements of the IG discourse continue to be the core part of the picture, but they do not exhaust the totality of the discussion in the domain. Social issues such as freedom of expression, human rights, digital divides, and, also, governance mechanism reforms are reaching prominent positions in the IG discourse and testify to the impossibility of treating ICTs and society separately. Neither of these two strands of the IG discourse is overwhelmingly prevailing over the other. Social and technical issues are equally relevant parts of ongoing conversations, and, as the operationalization of individual frames we pursued in this work has shown, they both imbue perceptions guiding participants' action in the domain. In this sense, it is interesting to notice that, although the Internet Governance Forum (IGF) process has helped to definitively consolidate the legitimacy of the IG social side, thus leading to an agenda enlargement, the challenge for the future seems not to be that of further broadening the thematic boundaries of the domain. Online and offline networks suggest, instead, that efforts should be pushed toward balancing opposite and contrasting needs, such as universal access, the security of users (all of them), and the management of ICRs. In this sense, a certain resistance to including new themes in the discussion (such as the relationship between IG and environmental concerns, which, so far, has had little appeal) can be interpreted in terms of a predominance of agenda refinement and systematization processes in the limitless expansion of the discussion, which, in the end, could jeopardize the effort of multiactor collaboration.

Online and offline semantic networks provided us also with useful insights on how this agenda refinement and systematization effort is taking place within a multiactor discursive context. All in all, it is definitely true that IG can be understood largely as including potentially everything regarding information and communication issues (Mathiason et al. 2004), but it is also true that not all issues are included in the IG discourse in the same way. Certainly, security and privacy are the two areas that more than any other have supplemented the technical flavor of the IG discourse both online and offline. This element is particularly meaningful in the case of online thematic networks, where possibly everything could become part of the discourse. The progressive channeling of the discourse around thematic clusters suggests that, after the inevitable widening of contents, a process of prioritization is occurring, aimed at collectively identifying the common core of IG matters around which other concerns and claims will be organized.

Beside commonalities and similarities among the two spaces, it is useful to look also at inconsistencies between the two discursive planes. One thematic cluster that is present online but not offline is the whole area of free software. This offline absence distinguishes the IGF from the World Summit on the Information Society (WSIS), where the free software community was actively and explicitly involved (especially in the Tunis phase). One possible interpretation could be that, within the IGF offline space, the free software movement melts within broader discourses on openness and access to knowledge without adopting a high and identifiable profile within the IGF process. However, it is also possible that this low profile is a consequence of the progressive cooling down of the countercultural mobilization efforts that emerged so powerfully during the WSIS process. In this sense, far from disappearing from the offline discussion, concerns brought by these and other radical groups that were involved in the WSIS continue to be articulated in the online space and can enter the IG discourse through broader and bridging themes, such as open and accessible knowledge, thus without finding in the official process a suitable space for strongly affirming their identity.

And yet, in spite of divergences like this one, there is no reason to expect that sooner or later the two discursive spaces will converge and totally overlap. The online and the offline, in the end, provide two different discursive planes whose potential can be exploited to different extents by different actors. The interesting (but still open!) question then concerns the reasons for which some issues or thematic clusters are developed mainly in one space or in the other, whereas some others find fertile ground for discussion in both spaces.

Coming to IG processes, our analyses showed that the collaborative multiactor dynamic supporting the construction of an IG discourse is still a work in progress and that there are different and intertwining trends contributing to its development. The exploration of online networks showed a high degree of fragmentation in communication patterns where thematic strands are deepened through exchanges of e-mails with low participation. Only in a few cases does online communication go beyond an informative function and show a real exchange of views and opinions: in most cases online direct communications serve to manage organizational tasks and only seldom constitute the vehicle for dealing with substantial thematic clarifications. This finding actually questions the effectiveness of IGF multiactor online spaces of participation. While Dynamic Coalitions (DCs) mailing list activity remains feeble, a lot of discussion is still going on in a highly participated way within the Internet Governance Caucus (IGC) established during the WSIS. On a general level, it seems that the multiplication of spaces for online direct communication between actors has led more to a fragmentation of communication than to its reinforcement.

Still, the fragmentation of online communication among Dynamic Coalitions (DCs) members that we pointed out through our analyses is not due to a higher weakness of online social relations in comparison to those established offline or to their residual possibility to inform the construction of an IG discourse. In fact, the IGC intense mailing list activity shows that a very lively discourse on IG can be carried on also online. However, the IGC and DCs mailing lists differ in their scope, as the latter is supposed to support multiactor discussion and the former to coordinate and feed a non-institutional , technical, and social discourse on IG. Rather, the current state of IGF-related mailing lists ties back to the idea that online communication, although potentially permitting the creation of bridges between different groups on the basis of shared contents and practices, in the end reinforces ties between like-minded individuals and groups (Norris 2004). In this sense, the difference between the IGC and DC mailing lists points to the difficulties of exploiting direct online communications to create bridging social capital, instead of strengthening ties among members of the same group (Putnam 2000). All in all, then, it seems that if multistakeholderism is hard to achieve in the real world, it is even harder to pursue it online. The boundless space created by the Internet seems to allow a minimalistic form of public interaction that, in all cases, takes the form of an information-sharing relation, rather than that of mutual dialogue. More in-depth and comparative studies would be needed to understand how it is possible to leverage on ICTs' communicative potential to enhance multiactor collaboration. Yet, it seems reasonable to hypothesize that a consolidation of offline recognition and trust mechanisms between traditional and nontraditional political actors will play a paramount role in this sense.

In this regard, the study of social ties in the offline world has shown that there are different logics guiding the construction of a shared partnership environment at the crossroads between bonding and bridging groups of actors. On the one hand, we found long-term solidarity mechanisms, for which collaborations in the IGF environment constitute the prosecution of collaborative relationships established during the WSIS. On the other hand, we found a more instrumental solidarity pattern, for which the IGF provides a first occasion to converge and collaborate on themes that constitute common concerns. Although both these logics reside within our offline social network, they have a different impact on the creation of a multiactor IG discourse. Indeed, individuals who already participated in the WSIS and did already deal with privacy and freedom of expression in that context are more central in the network of partner sharing and, thus, more central in the creation of the discourse, whereas individuals who are interested in openness and access to knowledge, although they are integrated in the discussion, are more peripheral. The IGF then seems to play a twofold function: it is a place for consolidating already existing strands of reflection and also a place for convergence, where old and new strands flood into it together, thus progressively structuring one in relation to the others.

Furthermore, the study of offline social ties also allows us to elaborate to a deeper extent on the roles of institutional and non-institutional actors within the construction of a common discourse about IG matters. Although actors of all statuses participate in the construction of the IG discourse on an equal footing, that is, enjoying the same set of opportunities for contributing to the construction and the consequent resolution of the IG social problem, they neither provide the same contribution to the process nor play the same function within conversational dynamics. While non-institutional actors play a major role in dispelling the IG uncertainty thanks to their high communicative potential and their provision of competencies and differentiated knowledge, institutions play a crucial role in establishing the conditions for an actual multistakeholder collaboration. In other words, without a full involvement of institutions in the discussion, the chances of actually achieving a shared and implementable frame on IG are actually jeopardized.

And yet, the more interesting result that emerged from our analysis is the importance of orientations and frames in guiding the construction of partnership relations. The relevance of inbred attributes, such as membership sector, is more limited in comparison to attitudes actors have toward the collective construction of the IG discourse. While reaching out to broader socially oriented issues amplifies the possibilities of constructing inclusive and participatory dynamics, it is the keenness for process building that determines a higher influence in the construction of the IG discourse, in terms of both establishing direct contacts and of contributing to the creation of favorable interactional environments. In general, then, it is the progressive construction and acceptance of new logics of mutual recognition, based on similarity and proximity reconceived in terms of understanding and political will rather than baseline qualities, that influence the possibility to successfully collaborate in the IG domain. Formal openness without explicit attempts to grasp the opportunities provided does not, overall, reform political arrangements.

The relevance of the IGF process in terms of procedural uncertainty reduction should be then assessed in relation to the redefinition of norms of political interaction. Social dialogue dynamics on IG seem to be developing in a space that is structured similarly to the ideal communicative situation described by Habermas (1986), in which equal opportunities for speaking are provided to all participants, equity of status is being granted by the definition of parity in dignity (which is different from the equality of roles), and the very goal of participation is enhancement of discussion according to a principle of cooperation (Grice 1999 [1975]). In such a situation, the ultimate arbiter should be, according to Habermas, the rationality of the argument proposed and not the identity of claimants. The IG case seems to suggest that, to a certain extent, traditional identity marks are less important than what actors are saying in the discussion. Far from functioning perfectly and being close to achieving a definitive redefinition of political roles, multiactor conversation is nonetheless leading to a progressive reduction of procedural uncertainty starting from open and participatory confrontation on political contents.

2. INTERNET GOVERNANCE, GLOBAL COMMUNICATION GOVERNANCE, AND THE REFORM OF WORLD POLITICS

While they help us uncover relevant mechanisms in the IG domain, results obtained by our network study of the IG domain also tell us something about how the GCG field is being structured, from both the content and the process points of view. Networks allow us to go beyond a study of multiactor dynamics centered on the sole category of the presence or absence of themes and actors on the scene. Indeed, if we lean only on this latter perspective, the thematic richness that emerged thanks to civil society involvement in the discussion during the WSIS process seems maintained in the passage to the IGF. The open IGF environment welcomes insights and inputs from all participants and, consistent with its premises, provides a location for homeless issues. But, if we look at how themes are organized within the IG discourse, although many thematic inputs "are there" in the IG agenda, only a few of them have become priorities (i.e., security, access, and ICRs), and it is around them that the rest of the IG thematic maze is developing. Certainly, the IG agenda is broader than before, and this fact suggests that an enlarged participation, fostered by the absence of constraints tied to status, actually broadens the scope of GCG and leads to the systematization of a complex discourse. In this context, as semantic networks have illustrated, the challenge is not to add further corollary elements to the already rich thematic framework but, rather, to find a balance between opposite needs that are equally standing at the core of the discussion.

The examination of contents here ties back to the matter of processes. If the thematic challenge is to find a balance between needs that are all legitimately entering the GCG field, how can it be faced through efficient political behavioral practices? Results obtained in particular from the study of offline collaboration networks suggest that the "mentality change" (Padovani 2005a) that is necessary to reform politics is actually happening. Partnerships are beginning to be established on the bases of frames and orientations, leaving status matters behind. However, the relevance of cognitive resources in defining partnerships is not tantamount to the removal of roles in the construction of a common discourse. Participatory settings seem to be leading to a *redefinition* of roles that, in the end, is not a total revolution in this case. Civil society entities provide the necessary knowledge for the management of information and communication issues, while governments remain crucial in setting up appropriate environments to receive and make this knowledge circulate. The level of success of multiactor dynamics is strictly connected to the diffusion of a collaborative attitude that should be adopted both by institutional and non-institutional participants. In other words, openness does not necessarily translate into innovation and change. Still, it remains a prerequisite on the way toward it.

However, if we look at the value of the IGF experience for the broader GCG field, a question on the levels of commitment of different actors remains open. The dynamics we have examined in this book are ultimately tied to the production of behavioral norms for multiactor relationships, so they stand one step behind the crystallization of formal rules. This element generates a sort of paradox that affects both the IGF and multiactor politics in general. On the one hand, the fact that multiactor dynamics are not intended to reach any binding results means that divergence in positions is accepted and not feared, as it was in the WSIS case. This facilitates the convergence of actors and issues, as the risk to get involved in the conversation is not perceived as too high. On the other hand, though, the impossibility of upping the ante within formal negotiation processes raises doubts about the level of commitment to the development of the discourse characterizing institutional (but also non-institutional) actors, who are still guaranteed a privileged access to policy-making processes. Are they collaborating in the construction of the discourse together with public interest groups only as a façade or, on the contrary, are they participating in something that they genuinely consider a mutual learning experience? In this latter case, how much of what is learned during the IGF experience is then imbuing a reformed and informed policy making? To answer this question, further research activities are needed and should be deployed according to a longitudinal and cross-sectional perspective, for which the networks wherein the political discourse is created are compared and considered in conjunction with the study of policy-making processes, thus assessing network impacts, potentials, and losses.

In any case, few would doubt that advancements in the IG and GCG fields toward an increased participation and toward more democratic arrangements are certainly creating a meaningful precedent for the reform of world politics. Yet, the two interrelated features of content and process as we examined them in this book seem to call for particular caution in generalizing results. We saw that, within the IG domain, the collaboration between institutional and non-institutional actors is leading toward a progressive redefinition of roles that are less tied to formal status and more to individual frames and cognitive understanding of the issue at stake. But how easy would it be to translate this mechanism into other domains of global politics?

In the first place, IG remains, despite its increased evidence, a rather unknown domain in comparison to other, more popular global issues, such as global justice and peace, environmental concerns, and so forth. This minority position does not relate solely to the popularity of the topic itself but also follows from the fact that mobilization in this field requires a specific expertise. Thus, expertise is not only technical in this domain but also basically tied to the ability to perceive, translate, and, ultimately, communicate to the broader public the consequences of the management of the Internet and of the virtual space. In this sense, the exportability of the IG experience seems to be tightly bound to the presence of intellectuals as interpreters (Baumann 1987) who have to engage in a twofold and hard effort: contributing to the development of a *lingua franca* that smooths the difficulties of multiactor dialogue within the IG domain and, at the same time, contributing to the construction of the IG social problem, starting from the discussion in the domain and "translating" it outside the debate forums to nonexpert, nontechnical, everyday Internet users, who should remain, in the end, the very point of reference for any reform of IG politics.

The work of intellectuals as translators à la Baumann is particularly crucial to overcome the obstacles posed by the non-materiality of communication resources (in our case of the Internet space) to the consolidation of IG and GCG as crucial global issues in the larger global scenario. The challenge of connecting the IG experience to other domains of global, regional, and local politics can be faced only if a tie is sent from what happens to us *on* the Internet and *through* the Internet (our Facebook friends, our e-mail activity, our information surfing, etc.) to what happens *to* the Internet itself. More generally, the capacity to piggyback on real events, which transformed the nuclear threat into a general concern (Ungar 1992), seems to be quite a difficult task in the IG and the GCG cases but, nonetheless, remains the key for consolidating the IG experiment on the global scale. In this perspective, more emphasis should be put on portraying the lack of availability, access, or competencies in terms of inequalities or information-poorness and as a violation of personal integrity. More than that, the participation of users in open discussions, the possibility of representing properly and universally different publics, their needs and exigencies, and the consideration of inputs provided within official processes such as IGF should be considered the mark of a new, participative global democracy. In this sense, they should be defended, in the IG domain as well as in all other domains of world politics, to achieve a genuinely representative global system.

3. NETWORKS AND WORLD POLITICS

In this book, we employed networks as our principal investigation tool, and we argued that it is through a relational approach that we can jointly study IG (and, more generally, GCG) contents and processes. Through the study of semantic and social networks we could gain some new insights on our case study but also on how world politics are changing. Along network ties we weighed general considerations on thematic richness and open participation, we examined their actual translation into the discursive environment for IG, and we elaborated on how this experience is contributing to a redefinition of the political roles of institutional and non-institutional actors in global politics.

Still, it should also be acknowledged that the adoption of a relational perspective implied some difficulties. Networks provided us with a flexible tool to analyze our objects of interest, but they had to be flexibly adapted to a general context where data are scant and hard to obtain and where interpersonal ties are difficult to trace. In the first place, we had to reduce the complexity that characterizes the transnational dynamics fostered by the IGF process. In our case, we started from DCs on IG (see Chapter 3), but this decision immediately challenged us on several points, such as the nonproportionality of stakeholder representation or the low presence of non-Northern, non-Western countries. Whether one agrees or disagrees with starting from DCs to investigate the discursive dynamics developing from the IGF context, the underlying problem remains: passing from a metaphorical to an empirical use of networks for investigating transnational dynamics entails starting from somewhere in time, space, and social and thematic relations. Every choice made, in this sense, will generate different consequences and biases, for the fluid nature of the IG discourse is hard to keep under control.

One obstacle to the full exploitation of the network potential in our case was the impossibility of studying partnership as a direct relationship between actors. Indeed, in our study we could study partnership-building processes only indirectly, given the impossibility of reaching all individuals defined by our interviewees as partners. More than that, we missed much information about active posters beyond the few clues provided by e-mail addresses or signatures; online thematic networks provided us with a somewhat static snapshot in the continuously evolving Web reality; and offline semantic networks should be complemented with a more in-depth analysis of the profound meanings that individuals attach to every thematic input they submit. And yet, as we are in an early stage of actually applying networks and network techniques to the study of global politics and the GCG field in particular, the weaknesses of this study should be considered as the departure point for improving along the way. All in all, the study of contemporary global discussions on IG and on information and communication further *confirms* for us that networks are powerful and flexible tools, both theoretically and empirically, that nonetheless require some caution in the interpretation and generalization of results, especially when data are difficult to obtain. The exploitation of their potential requires attention and a great amount of work.

And yet, if networks are reconstructed and interpreted while acknowledging the holes and weaknesses deriving from the many difficulties that arise along the way, their heuristic potential is enormous. The IG case showed that, when contemporary changes in world politics are studied beyond the surface, it is necessary to keep an eye simultaneously on "who you are" and "what you think." Networks are fundamental tools in this regard because they account simultaneously for the contents and processes developed by actors in the field. But besides providing the starting point for examining what IG is progressively becoming, networks also allow us to see "what is not out there," at least for now. Where network ties are present, some kind of potential is being exploited. Where network ties are absent, some kind of potential is inactive, but it is still there. Networks help us go beyond the perennial game of opposites because they show, at the same time, what results are achieved and what are not achieved but also, most importantly for the reform of political mechanisms, where action could be undertaken to improve the overall effort to move global politics toward a radical and genuine democracy.

Appendix

Issue Crawler Working Logic

The Issue Crawler is a tool for locating and visualizing online thematic networks that has been developed by the Govcom.org Foundation of Amsterdam.[1] Thematic networks are just one of the types of networks that the software can trace and visualize;[2] they are aimed at identifying which set of Web resources are talking, in our case, about Internet governance (IG) on the Web. In order to do so, the software needs a list of *starting points*, namely, of URLs from which its activity of locating a thematic network will start. Identification of starting points is a crucial phase in the research path because it influences the directions that the network will follow in its development. There are several ways to identify a list of starting points. One might simply query a search engine for the issue area of interest. In our case, we could have gone into the Google.com page and asked for "Internet governance." Then we might have chosen some of the results proposed, let's say the first 20 URLs in the list, as our starting point. This implies, of course, that the list we are adopting is based on peculiar search engine logic and also presupposes our familiarity with the ranking principles of the engine itself, which we need to evaluate the results of the search (Rogers 2006, 2008). A second way might be the most classical method for mapping network boundaries: asking "experts" in the area for the names of the most relevant organizations in the field. Once we had obtained organization names, we would look for the corresponding URLs online and use that list as our departure point. In some other cases, such as in our specific case, the researcher might have some idea of relevant URLs from which the search can be started.

As previously mentioned, online thematic networks are based but do not coincide with links sent from one site to others. The algorithm through which the software works is based on two main operations. In the first place, Issue

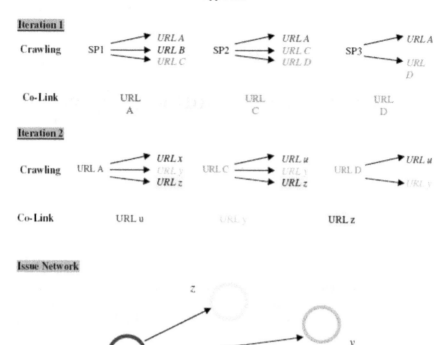

Figure A.1.

Crawler goes to each URL defined as a *starting point* and fetches all outlinks present in that page. This operation is called *crawling*. Often, instead of providing the software with home page URLs, we can choose to submit as a starting point the *Links* or *Resources* pages, as these are specifically dedicated to host all ties to other websites. Issue Crawler looks for outgoing links from each URL provided as a *starting point* and moves to perform its key step, the *co-link analysis*. According to this method, after fetching all outgoing links, the Issue Crawler retains only those URLs that have been commonly identified as outlinks by at least two *starting points*. Those outgoing links identified by only one *starting point* are not considered. It is possible to decide how many times crawling and co-link analysis have to be repeated (an *iterations number* that ranges from 1 to 3). If iterations are set >1, then URLs identified through the co-link analysis will become new *starting points* and the next crawling and co-link analysis operations will be conducted on that list.[3] Thus, it is possible to decide the *depth* of levels at which sites are considered (from 1 to 3).[4] Analyses can be performed either by page (when the co-link analysis is performed on deep pages) or by site (when the co-link analysis is performed on the hosts and not on deep pages).[5] After specifying all these settings, the Issue Crawler starts its functioning and stops after the number of iterations specified. After crawling and co-link analysis are repeated as many times as required, the software retains a set of final co-links and additionally looks for interlinks existing among them (see Figure A.1). Therefore, the final network will be composed of nodes identified in the last co-link analysis step and of the ties existing between them. Results are memorized in an .xml file that is then visualized as a cluster map (like the one shown in Figure 4.2 in Chapter 4), a circle map (where all nodes are visualized in a circle), or a world map on which URLs are place depending on places in which each URL is registered.

NOTES

1. www.govcom.org and www.issuecralwer.net.

2. Other types are personal networks (immediate neighborhood on the Web), establishment networks (institutional thematic environment on the Web); event networks (who goes to which event); and network evolution over time (through the scheduler function, which allows one to retrace the same network automatically over time).

3. If users wish to do so, they might add their *starting points* as seeds for the second iteration, together with the set of co-linkees, through the option "*Privilege starting points ON.*"

4. "The pages fetched from the starting point URLs are considered to be depth 0. The pages fetched from URL links from those pages are considered to be depth 1. In general, the pages found from URL links on a page of depth N are considered to be depth N+1. If you set a depth of 2, then no pages of depth 2 will be fetched. Only pages of depth 0 and 1 will be fetched (i.e., two levels of depth)" (source: www .govcom.org/Issuecrawler_instructions.htm).

5. A host is defined as a source_host (e.g., www.example.com), while a deep page goes more in depth as far as URL levels go (e.g., www.example.com/ex/am/ple/index.html) (source: www.govcom.org/Issuecrawler_instructions.htm).

Bibliography

PUBLICATIONS

Abbott, Andrew. 1995. "Things of Boundaries." *Social Research* 62(4): 857–882.

Adam, Lishan, Tina James, and Alice Munya Wahjira. 2007. *Frequently Asked Questions about Multi-Stakeholder Partnerships in ICTs for Development: A Guide for National ICT Policy Animators.* Association for Progressive Communication (APC) www.apc.org/en/system/files/catia_ms_guide_EN-1.pdf.

Adam, Silke, and Hanspeter Kriesi. 2007. "The Network Approach." In *Theories of the Policy Process*, ed. Paul A. Sabatier, 129–154. Boulder, Colo.: Westview Press.

Adler, Emanuel. 1997. "Seizing the Middle Ground: Constructivism in World Politics." *European Journal of International Relations* 3: 319–363.

Alger, Chadwick F. 1997. "Transnational Social Movements, World Politics and Global Governance." In *Transnational Social Movements and Global Politics: Solidarity beyond the State*, ed. Jackie Smith, Charles Chatfield, and Ron Pagnucco, 260–278. New York: Syracuse University Press.

Anheier, Helmut, and Hagai Katz. 2004. "Network Approaches to Global Civil Society." In *Global Civil Society 2004/5*, ed. Helmut Anheier, Marlies Glasiusand, and Mary Kaldor, 206–221. London, Thousand Oaks, New Delhi: Sage Publications.

Ayres, Jeffrey M. 1999. "From the Streets to the Internet: The Cyber-Diffusion of Contention." *Annals of the American Academy of Political and Social Science* 566(1): 132–143.

Baiocchi, Gianpaolo. 2003. "Participation, Activism, and Politics: The Porto Alegre Experiment." In *Deepening Democracy: Institutional Innovation in Empowered Participatory Governance*, ed. Archon Fung and Eric Olin Wright, 45–76. London, New York: Verso.

Baird, Zoë, and Stefaan Verhulst. 2004. "A New Model for Global Internet Governance." In *Internet Governance: A Grand Collaboration*, ed. Don MacLean, 58–64. New York: United Nations ICT Task Force.

Baumann, Zygmunt. 1987. *Legislators and Interpreters: On Modernity, Post-modernity and Intellectuals.* Ithaca, N.Y.: Cornell University Press.

Bendrath, Ralf. 2005. "Civil Society and Multi-Stakeholderism." Accessed December 2007. http://www.worldsummit2005.org.

Benford, Robert. 1993. "Frame Disputes within the Nuclear Disarmament Movement." *Social Forces* 71(3): 677–701.

Benford, Robert D., and David A. Snow. 2000. "Framing Processes and Social Movements: An Overview Assessment." *Annual Review of Sociology* 26: 611–639.

Bijker, Wiebe E. 2006. "Why and How Technology Matters." In *The Oxford Handbook of Contextual Political Analysis*, ed. Robert E. Goodin and Charles Tilly, 681–706. Oxford: Oxford University Press.

Blumer, Herbert. 1971. "Social Problems as Collective Behavior." *Social Problems* 18(3): 298–306.

Boase, Jeffrey, John B. Horrigan, Barry Wellman, and Lee Rainie. 2006. "The Strength of Internet Ties." Accessed June 2010. http://homes.chass.utoronto.ca/~wellman/publications/.

Boissevain, Jeremy. 1974. *Friends of Friends.* Oxford: Basil Blackwell.

Borgatti, Steve P. Martin, G. Everett, and Linton C. Freeman, 2002. Ucinet for Windows: Software for Social Network Analysis. Harward, MA: Analytic Technologies.

Börzel, Tanja A. 1997. "What's So Special about Policy Networks? An Exploration of the Concept and Its Usefulness in Studying European Governance." Accessed March 2008. http://eiop.or.at/eiop/texte/1997–016a.htm.

———. 1998. "Organizing Babylon—On the Different Conceptions of Policy Networks." *Public Administration* 76: 253–273.

Botzem, Sebastian, and Jeanette Hofmann. 2008. *Transnational Institution Building as Public-Private Interaction: The Case of Standard Setting on the Internet and in Corporate Financial Reporting.* Discussion paper 51. London: London School of Economics and Political Science. http://eprints.lse.ac.uk/36535/1/Disspaper51.pdf.

Braman, Sandra. 2006. *Change of State: Information, Policy and Power.* Cambridge, Mass.: MIT Press.

Breiger, Ronald L. 1974. "The Duality of Persons and Groups." *Social Forces* 53(2): 181–190.

Bunting, David, and Jeffrey Barbour. 2002. "Interlocking Directorates in Large American Corporations, 1896–1964." In *Social Networks: Critical Concepts in Sociology* (vol. III), ed. J. Scott, 183–200. London, New York: Routledge.

Busaniche, Beatriz. 2006. "La società civile sulla giostra: Chi vince, chi perde e chi è dimenticato nell'approccio multistakeholder?" In *Saperi del futuro: Analisi di donne sulla società della comunicazione*, ed. Olga Drossou, Heike Jensen, and Claudia Padovani, 90–96. Bologna: Editrice Missionaria Italiana (EMI).

Caiani, Manuela, and Claudius Wangemann. 2009. "Online Networks of the Italian and German Extreme Right." *Information, Communication and Society*,12(1): 66–109.

Calabrese, Andrew. 2004. "The Promise of Civil Society: A Global Movement on Communication Rights." *Continuum: Journal of Media and Society* 18(3): 317–319.

Calderaro, Andrea. 2008. "Empirical Analysis of Political Spaces on the Internet: The Role of Mailing Lists in the Organization of the Global Justice Movement." Paper presented at the 58th ICA Conference, Montreal.

Calhoun, Craig. 1998. "Community without Propinquity Revisited: Communication Technology and the Transformation of the Urban Public Sphere." *Sociological Inquiry* 68(3): 373–397.

Cammaerts, Bart, and Nico Carpentier. 2004. "The Unbearable Lightness of Full Participation in a Global Context: WSIS and Civil Society Participation." In *Towards a Sustainable Information Society beyond WSIS*, ed. Jan Servaes and Nico Carpentier, 17–49. Bristol and Portland: Intellect.

Cammaerts, Bart, and Claudia Padovani. 2006. "Theoretical Reflections on Multi-Stakeholderism in Global Policy Processes: The WSIS as a Learning Space." Paper presented at the annual conference of the International Association for Media and Communication Research (IAMCR), Il Cairo.

Carley, Kathleen M. 1997. "Network Text Analysis: The Network Position of Concepts." In *Text Analysis for the Social Sciences*, ed. Carl W. Roberts, 79–100. Mahwah, N.J.: Lawrence Erlbaum Associates.

Carlsson, Ulla. 2003. "The Rise and Fall of NWICO—and Then? From a Vision of International Regulation to a Reality of Multilevel Governance." *Nordicom Review* 2: 31–67.

———. 2007. "From NWICO to Global Governance of the Information Society." In *Media e Glocal Change: Rethinking Communication for Development*, ed. Oscar Herne and Thomas Tufte, 193–214. Buenos Aires: CLACSO.

Castells, Manuel. 1996. *The Rise of Network Society*. Oxford: Blackwell Publishers.

———. 2000. "Materials for an Exploratory Theory of the Network Society." *British Journal of Sociology* 51(1): 5–24.

Cerulo, Karen. 1997. "Reframing Sociological Concepts for a Brave New (Virtual) World?" *Sociological Inquiry* 67(1): 48–58.

Cerulo, Karen, and Janet M. Ruane. 1998. "Coming Together: New Taxonomies for the Analysis of Social Relations." *Sociological Inquiry* 68(3): 398–425.

Chilton, Paul A. 2004. *Analysing Political Discourse: Theory and Practice*. London: Routledge.

Christopoulos, Dimitrios C. 2006. "Relational Attributes of Political Entrepreneurs: A Network Perspective." *Journal of European Public Policy* 13(5): 757–778.

CONGO and NGLS. 2005. "Orientation Kit." *United Nations Non-Governmental Liaison Service*. Accessed August 2010. www.un-ngls.org/orf/pdf/wsis-Orientation%20Kit%20-%20new%20version%20Rev.pdf.

Contractor, Noshir, Peter Monge and Paul Leonardi. 2011. "Multidimensional Networks and the Dynamics of Sociomateriality: Bringing Technology inside the Network." *International Journal of Communication* 5, 682–720.

Coombs, Clyde H. 1964. *A Theory of Data*. New York: Wiley.

Currie, Willie. 2005. "WSIS Update: APC Involvement in the Task Force on Financial Mechanisms." Accessed August 2010. http://www.apc.org/es/news/hr/world/wsis-update-apc-involvement-task-force-financial-m.

D'Arcy, Jean.1977. "Direct Broadcast Satellite and the Right of Man to Communicate." In *Right to Communicate: Collected Papers,* ed. L. S. Harms, Jim Richstad, and Kathleen Kie, 1–9. Honolulu: Social Sciences Institute. Distributed by the University Press of Hawaii. www.righttocommunicate.org/viewDocument.atm? sectionName=summaries&id=16.

Davis, Allison B., Burleigh B. Gardner, and Mary R. Gardner. 1959. *Deep South.* Chicago: University of Chicago Press.

de la Chapelle, Bertrand. 2007. "The Internet Governance Forum: How a United Nations Summit Produced a New Governance Paradigm for the Internet Age." In *Governing the Internet: Freedom and Regulation in the OSCE Region,* ed. Christian Möeller and Arnaud Amoroux, 19–28. OSCE Representative on Freedom of the Media.

———. 2010. "Towards an Internet Governance Network: Why the Format of the IGF Is One of Its Major Outcomes." In *Internet Governance: Creating Opportunities for All,* ed. William J. Drake, 92–106. New York: United Nations Publications.

Della Porta, Donatella, Massimiliano Andreatta, Lorenzo Mosca, and Herbert Reiter. 2006. *Globalization from Below: Transnational Activists and Protest Networks.* Minneapolis: University of Minnesota Press.

Di Maggio, Paul, and Eszter Hargittai. 2001. *From the "Digital Divide" to "Digital Inequality": Studying Internet Use as Penetration Increases.* Working Paper 47. Princeton, N.J.: Center for Arts and Cultural Policy Studies, Woodrow Wilson School, Princeton University. www.princeton.edu/~artspol/workpap/WP15%20-% 20DiMaggio%2BHargittai.pdf.

Di Maggio, Paul J., and Walter W. Powell. 1983. "The Iron Cage Revisited: Institutional Isomorphism and Collective Rationality in Organizational Fields." *American Sociological Review* 48(2): 147–160.

Diani, Mario. 1995. *Green Networks: A Structural Analysis of the Italian Environmental Movement.* Edinburgh: Edinburgh University Press.

———. 2000. "Comunità virtuali, comunità reali e azione collettiva." *Rassegna Italiana di Sociologia* 41(1): 29–52.

———. 2003. "Leaders or Brokers? Positions and Influence in Social Movements Networks." In *Social Movements and Networks: Relational Approaches to Collective Action,* ed. Mario Diani and Doug McAdam, 105–122. Oxford: Oxford University Press.

———. 2008. "Modelli di azione collettiva: Quale specificità per i movimenti sociali?" *Partecipazione e Conflitto,* no.1: 43–66.

———. 2009. "Network Structures of Collective Action." Paper presented at the conference "The Unexpected Link: Using Network Science to Tackle Social Problems," Centre for Network Science (CEU), Budapest.

Diani, Mario, and Ivano Bison. 2004. "Organizations, Coalitions and Movements." *Theory and Society* 33(3–4): 281–309.

Diesner, Jana, and Kathleen M. Carley. 2005. "Revealing Social Structure from Texts: Meta-Matrix Text Analysis as a Novel Method for Network Text Analysis." In *Causal Mapping for Information Systems and Technology Research:*

Approaches, Advances, and Illustrations, ed. V. K. Naraynan and Deborah J. Armstrong, 81–108. Harrisburg, Pa.: Idea Publishing Group.

Dodds, Felix. 2002. "The Context: Multi-Stakeholder Processes and Global Governance." In *Multi-Stakeholder Processes for Governance and Sustainability: Beyond Deadlock and Conflict*, ed. Minu Hemmati, 26–38. London: Earthscan.

Donati, Paolo. 1992. "Political Discourse Analysis." In *Studying Collective Action*, ed. Mario Diani and Ron Eyerman, 136–167. London: Sage Publications.

Drake, William J. 2004. "Reframing Internet Governance Discourse: Fifteen Baselines Preposition." In *Internet Governance: A Grand Collaboration*, ed. Don MacLean, 122–161. New York: United Nations ICT Task Force.

———. 2005. "Why the WGIG Process Mattered." In *Reforming Internet Governance: Perspectives from the Working Group on Internet Governance*, ed. William J. Drake, 249–265. New York: United Nations ICT Task Force.

Drake, William J., and Ernest J. Wilson III (eds.). 2008. *Governing Global Electronic Networks: International Perspectives on Policy and Power*. Cambridge, Mass.: MIT Press.

Dryzek, John S. 2005. "Deliberative Democracy in Divided Societies: Alternatives to Agonism and Analgesia." *Political Theory* 33(2): 218–242.

Esterhuysen, Anriette. 2005. "Multi-Stakeholder Participation and ICTs Policy Processes." Accessed December 2007. www.apc.org/en/news/access/world/multi-stakeholder-participation-and-ict-policy-pro.

———. 2008. "Reflections on the Internet Governance Forum from 2006–8." In *Internet Governance Forum (IGF): The First Two Years*, ed. Avri Doria and Wolfgang Kleinwächter, 37–41. UNESCO Publications.

Faist, Thomas. 2004. "The Border-Crossing Expansion of Social Space: Concepts, Questions and Topics." In *Transnational Social Spaces: Agents, Networks and Institutions*, ed. Thomas Faist and Eyup Őzveren, 1–36. Aldershot: Ashgate.

Faust, Katherine. 1997. "Centrality in Affiliation Networks." *Social Networks* 19: 157–191.

Fernandez, Roberto M., and Roger V. Gould. 1989. "Structures of Mediation: A Formal Approach to Brokerage in Transaction Networks." *Sociological Methodology* 19: 89–126.

———. 1994. "A Dilemma of State Power: Brokerage and Influence in the National Health Policy Domain." *American Journal of Sociology* 99(6): 1455–1491.

Finkelstein, Lawrence S. 1995. "What Is Global Governance?" *Global Governance* 1(3): 367–372.

Finnemore, Martha, and Kathryn Sikkink. 1998. "International Norm Dynamics and Political Change." *International Organization* 52(4): 887–917.

Fisher, Kimberly. 1997. "Locating Frames in the Discursive Universe." *Sociological Research Online* 2(3). www.socresonline.org.uk/socresonline/2/3/4.html.

Fraser, Nancy. 2005. "Transnationalizing the Public Sphere." *Republicart.net*. Accessed April 2007. http://www.republicart.net/disc/publicum/fraser01_en.htm.

Freeman, Linton C. 2002 [1979] "Centrality in Social Networks: Conceptual Clarifications." In *Social Networks: Critical Concepts in Sociology* (vol. I), ed. John Scott, 238–263. London and New York: Routledge.

Friedkin, Noah E. 1991. "Theoretical Foundations for Centrality Measures." *American Journal of Sociology* 96(6): 1478–1504.

Gallagher, Margaret. 1986. "Women and NWICO." In *Communication for All*, ed. Philip Lee, 33–56. New York: Orbis Book.

Galtung, Johan. 1986. "Social Communication and Global Problems." In *Communication for All*, ed. Philip Lee, 1–16. New York: Orbis Book.

Gamson, William A. 1988. "Political Discourse and Collective Action." In *International Social Movement Research, Vol. I: From Structure to Action: Comparing Social Movement Research across Cultures*, ed. Bert Klandermans, Hanspeter Kriesi, and Sidney Tarrow, 219–244. Greenwich, London: Jai Press.

———. 1992. *Talking Politics*. Cambridge: Cambridge University Press.

Gamson, William, and Andre Modigliani. 1989. "Media Discourse and Public Opinion on Nuclear Power: A Constructionist Approach." *American Journal of Sociology* 95(1): 1–37.

Goffman, Ervin. 1974. *Frame Analysis*. London: Penguin Books.

Grice, Herbert P. (1999 [1975]). "Logic and Conversation," pp. 66–77 in Adam Jaworski and Nicolas Couplord (eds.), *The Discourse Reader*, London: Routledge.

Habermas, Jurgen. 1986. *Teoria dell'agire comunicativo*. Bologna: Il Mulino.

———. 1989. *The Structural Transformation of the Public Sphere: An Inquiry into a Category of Bourgeois Society*. Cambridge, Mass.: MIT Press.

Hamelink, Cees J. 1994. *The Politics of World Communication—A Human Rights Perspective*. London: Sage.

Hanf, Kenneth, and Laurence J. O'Toole. 1992. "Revisiting Old Friends: Networks, Implementation Structures and the Management of Interorganizational Relations." *European Journal of Political Research* 21(1–2): 163–180.

Hanneman, Robert A., and Mark Riddle, 2005. *Introduction to Social Network Methods*. Riverside, CA: University of California, Riverside. Published in digital form at faculty.vcr.edu/hanneman.

Haythornthwaite, Caroline, and Berry Wellman. 2002. "The Internet in Everyday Life. An Introduction." In *The Internet in Everyday Life*, ed. Caroline Haythornthwaite and Berry Wellman, 3–41. Oxford: Blackwell Publishing.

Held, David, Anthony McGrew, David Goldblatt, and Johnatan Perraton. 1999. *Global Transformations: Politics, Economics, and Culture*. Cambridge: Polity Press.

Hemmati, Minu (ed.). 2002. *Multi-Stakeholder Processes for Governance and Sustainability: Beyond Deadlock and Conflict*. London: Earthscan.

Hewson, Martin, and Timothy J. Sinclair. 1999. "The Emergence of Global Governance Theory." In *Approaches to Global Governance Theory*, ed. Martin Hewson and Timothy J. Sinclair, 2–22. Albany: State University of New York Press.

Hilgartner, Stephen, and Charles L. Bosk. 1988. "The Rise and Fall of Social Problems: A Public Arenas Model." *American Journal of Sociology*, 94(1): 53–78.

Hintz, Arne. 2007. "Civil Society Media at the WSIS: A New Actor in Global Communication Governance?" In *Reclaiming the Media*, ed. Bart Cammaerts and Nico Carpentier, 243–264. Bristol: Intellect.

———. 2009. *Civil Society Media and Global Governance: Intervening into the World Summit on the Information Society*. Münster: LIT Verlag.

Hockings, Brian. 2006. "Multistakeholder Diplomacy: Forms, Functions and Frustrations." In *Multistakeholder Diplomacy: Challenges and Opportunities*, ed. Jovan Kurbaljia and Valentin Katrandjiev, 13–32. Geneva and Malta: Diplo Foundation.

Hofmann, Jeanette. 2006. *Internet Governance: A Regulative Idea in Flux.* Accessed August 2007. http://duplox.wzberlin.de/people/jeanette/texte/Internet%20Governance%20english%20version.pdf.

Jepperson, Ronald L. 2000. "Istituzioni, effetti istituzionali e istituzionalismo." In *Il neoistituzionalismo nell'analisi organizzativa*, ed. Walter W. Powell and Paul J. Di Maggio, 195–222. Torino: Edizioni Comunità.

Johnston, Hank. 2002. "Verification and Proof in Frame and Discourse Analysis." In *Methods of Social Movement Research,* ed. Bert Klandermans and Suzanne Staggenborg, 62–91. Minneapolis, London: University of Minnesota Press.

Jones, Candance, William S. Hersterly, and Steve P. Borgatti. 1997. "A General Theory of Network Governance: Exchange Conditions and Social Mechanisms." *Academy of Management Review* 22(4): 911–945.

Kahler, Miles. 2009. "Networked Politics: Agency, Power and Governance." In *Networked Politics: Agency, Power and Governance*, ed. Miles Khaler, 1–20. Ithaca, N.Y., and London: Cornell University Press, 2009.

Katz, Hagai, and Helmut Anheier. 2006. "Global Connectedness: The Structure of Transnational NGO Networks." In *Global Civil Society 2005/6*, ed. Marlies Glasius, Mary Kaldor, and Helmut Anheier, 240–265. London: Sage.

Katz, James E., and Ronald E. Rice. 2002. *Social Consequences of Internet Use: Access, Involvement and Interaction.* Cambridge, Mass.: MIT Press.

Kenis, Patrik, and Volker Schneider. 1991. "Policy Network and Policy Analysis: Scrutinizing a New Analytical Toolbox." In *Policy Networks: Empirical Evidence and Theoretical Considerations*, ed. Bernd Marin and Renate Mayntz, 25–62. Boulder, Colo.: Westview Press.

Keohane, Robert O., and Joseph Nye. 1977. *Power and Independence: World Politics in Transition.* Boston: Little, Brown.

Khagram, Sanjeev, James V. Riker, and Kathryn Sikkink. 2002. "From Santiago to Seattle: Transnational Advocacy Groups Restructuring World Politics." In *Restructuring World Politics: Transnational Social Movements, Networks and Norms*, ed. Sanjeev Khagram, James V. Riker, and Kathryn Sikkink, 3–23. Minneapolis: University of Minnesota Press.

Kidd, Dorothy, and Clemencia Rodriguez. 2009. "Introduction." In *Making Our Media: Global Initiatives toward a Democratic Public Sphere* (vol. 1), ed. Laura Stein, Dorothy Kidd, and Clemencia Rodriguez, 1–22. Cresskill, N.J.: Hampton Press.

Klein, Hans. 2004. "Understanding WSIS: An Institutional Analysis of the UN World Summit on the Information Society." *Information Technologies and International Development* 1(3–4): 3–13.

Kleinwächter, Wolfgang. 2004. "Beyond ICANN vs. ITU? How WSIS Tries to Enter the New Territory of Internet Governance." *Gazette: The International Journal of Communication* 66(3–4): 233–251.

———. 2007. "The History of Internet Governance." In *Governing the Internet*, ed. Cristian Mőeller and Arnaud Amoroux, 41–66. OSCE Representative on Freedom of the Media.

———. 2010. "Multistakeholderism at the IGF: Laboratory, Clearinghouse, Watchdog." In *Internet Governance: Creating Opportunities for All*, ed. William J. Drake, 76–91. New York: United Nations Publications.

Knoke, David. 1990. *Political Networks: The Structural Perspective*. Cambridge: Cambridge University Press.

Knoke, David, and Ronald Burt. 1983. "Prominence." In *Applied Network Analysis: A Methodological Introduction*, ed. Roland S. Burt and Michael J. Minor, 195–222. Beverly Hills, London, New Delhi: Sage Publications.

Knoke, David, and James H. Kuklinski. 1982. *Network Analysis*. London: Sage Publications.

Knoke, David, Franz U. Pappi, Jeffrey Broadbent, and Yutaka Tsujinaka. 1996. *Comparing Policy Networks*. Cambridge: Cambridge University Press.

Kooiman, Jan. 2003. *Governing as Governance*. London, Thousand Oaks, New Delhi: Sage Publications.

Koopmans, Ruud. 2004a. "Political. Opportunity. Structure. Some Splitting to Balance the Lumping." In *Rethinking Social Movements: Structure, Meaning and Emotions*, ed. Jeff Goodwin and James M. Jasper, 61–74. Lanham, Md.: Rowman and Littlefield.

———. 2004b."Protests in Time and Space: The Evolution of Waves of Contention." In *The Blackwell Companion to Social Movements*, ed. David A. Snow, Sarah A. Soule, and Hanspeter Kriesi, 19–46. Oxford: Blackwell Publishing.

Kriesi, Hanspeter. 2004. "Political Context and Opportunity." In *The Blackwell Companion to Social Movements*, ed. David A. Snow, Sarah A. Soule, and Hanspeter Kriesi, 67–90. Oxford: Blackwell Publishing.

Kummer, Markus. 2004. "The Results of the WSIS Negotiations on Internet Governance." In *Internet Governance: A Grand Collaboration*, ed. Don MacLean, 53–57. New York: United Nations ICT Task Force.

———. 2010. "The Third IGF Book." In *Internet Governance: Creating Opportunities for All*, ed. William J. Drake, i–iii. New York: United Nations Publications.

Kurbalija, Jovan. 2008. *Internet Governance: An Introduction*. Geneva and Malta: Diplo Foundation.

Latham, Robert. 1999. "Politics in a Floating World." In *Approaches to Global Governance Theory*, ed. Martin Hewson and Timothy J. Sinclair, 23–54. Albany: State University of New York Press.

Latham, Robert, and Saskia Sassen. 2005. "Digital Formations: Constructing an Object of Study." In *Digital Formations: Information Technology and New Architecture in the Global Realm*, ed. Robert Latham and Saskia Sassen, 1–34. Princeton, NJ: Princeton University Press.

Laumann, Edward O., and John P. Hintz. 1991. "Organizations in Political Action: Representing Interests in National Policy Making." In *Policy Networks. Empirical Evidence and Theoretical Considerations*, ed. Bernd Marin and Renate Mayntz, 63–96. Boulder, CO: Westview Press.

Laumann, Edward O., and David Knoke. 1987. *The Organizational State*. Madison: University of Wisconsin Press.

Laumann, Edward O., Peter V. Marsden, and David Prensky. 1983. "The Boundary Specification Problem." In *Applied Network Analysis: A Methodological Introduction*, ed. Roland. S. Burt and Michael. J. Minor, 18–34. Beverly Hills, London, New Delhi: Sage Publications.

Lee, Micky. 2003. "A Historical Account of Critical Views on Communication Technologies in the Context of NWICO and the MacBride Report." Paper presented at the Euricom Colloquium, Padova-Venice.

Lorrain, François, and Harrison C. White. 1971. "Structural Equivalence of Individuals in Social Networks." *Journal of Mathematical Sociology* 1(1): 49–80.

MacLean, Don. 2004a. "Introduction." In *Internet Governance: A Grand Collaboration*, ed. Don MacLean, 1–10. New York: United Nations ICT Task Force.

———. 2004b. "Herding Schrödinger's Cats: Some Conceptual Tools for Thinking about Internet Governance." In *Internet Governance: A Grand Collaboration*, ed. Don MacLean, 73–99. New York: United Nations ICT Task Force.

Malcom, Jeremy. 2008. *Multi-Stakeholder Governance and the Internet Governance Forum*. Wembley: Terminus Press.

Marin, Alexandra, and Berry Wellman. 2010. "Social Network Analysis: An Introduction." http://homes.chass.utoronto.ca/~wellman/publications.

Marin, Bernd, and Renate Mayntz. 1991. "Introduction: Studying Policy Networks." In *Policy Networks: Empirical Evidence and Theoretical Considerations*, ed. Bernd Marin and Renate Mayntz, 11–24. Boulder, CO: Westview Press.

Mariolis, Peter. 1975. "Interlocking Directorates and Control of Corporations: The Theory of Bank Control." *Sociological Quarterly* 56: 425–439.

Marsh, David, and Martin Smith. 2000. "Understanding Policy Networks: Towards a Dialectical Approach." *Political Studies* 48(1): 4–21.

Martin, John L. 2003. "What Is Field Theory?" *American Journal of Sociology* 109(1): 1–49.

Mathiason, John. 2008. *Internet Governance: The New Frontier of Global Institutions*. London: Routledge.

Mathiason, John, Milton Mueller, Hans Klein, Marc Holitscher, and Lee McKnight. 2004. "Internet Governance: The State of Play." www.internetgovernance.org/pdf/ig-sop-final.pdf.

McAdam, Doug. 1982. *Political Process and the Development of Black Insurgency: 1930–1970*. Chicago: University of Chicago Press.

———. 2003. "Beyond Structural Analysis: Toward a New Dynamics Understanding of Social Movements." In *Social Movements and Networks: Relational Approaches to Collective Action*, ed. Mario Diani and Doug McAdam, 281–298. Oxford: Oxford University Press.

McPherson, Miller, Lynn Smith-Lovin, and James M. Cook. 2001. "Birds of a Feather—Homophily in Social Networks." *Annual Review of Sociology* 27: 415–444.

Melo, Marcus A., and Gianpaolo Baiocchi. 2006. "Deliberative Democracy and Local Governance: Towards a New Agenda." *International Journal of Urban and Regional Research* 30(3): 587–600.

Melucci, Alberto. 1984. *Altri codici: Aree di movimento nella metropolis*. Bologna: Il Mulino.

———. 1988. "Getting Involved: Identity and Mobilization in Social Movements." In *International Social Movement Research, Vol. I: From Structure to Action: Comparing Social Movement Research across Cultures*, ed. Bert Klandermans, Hanspeter Kriesi, and Sidney Tarrow, 329–350. Greenwich, London: Jai Press, 1988.

———. 1996. *Challenging Codes: Collective Action in the Information Age*. Cambridge: Cambridge University Press.

Meyer, David S. 2004. "Tending the Vineyard: Cultivating Political Process Research." In *Rethinking Social Movements: Structure, Meaning and Emotions*, ed. Jeff Goodwin and James M. Jasper, 47–60. Lanham, MD: Rowman and Littlefield.

Meyer, John W., and Brian Rowan. 2000. "Le organizzazioni istituzionalizzate: La struttura formale come mito e cerimonia." In *Il neoistituzionalismo nell'analisi organizzativa*, ed. Walter W. Powell and Paul J. Di Maggio, 59–87. Torino: Edizioni Comunità.

Milan, Stefania. 2006. "Democrazia/e e democrazia/e della comunicazione: Mediattivismo tra esperimenti di emancipazione e campagne di riforma." *Rivista Italiana di Sociologia* 47(4): 557–582.

Mische, Ann. 2003. "Cross-Talk in Movements: Reconceiving the Culture-Network Link." In *Social Movements and Networks: Relational Approaches to Collective Action*, ed. Mario Diani and Doug McAdam, 258–280. Oxford: Oxford University Press.

Monge, Peter R., and Noshir S. Contractor. 2003. *Theories of Communication Networks*. Oxford: Oxford University Press.

Mossberger, Karen, Caroline J. Tolbert, and Mary Stansbury. 2003. *Virtual Inequalities: Beyond the Digital Divide*. Washington DC: Georgetown University Press.

Mowlana, Hamid. 1993. "Toward a NWICO for the Twenty-First Century?" *Journal of International Affairs* 47(1): 59–72.

Mueller, Milton L. 2002. *Ruling the Root: Internet Governance and the Taming of Cyberspace*. Cambridge, London: MIT Press.

———. 2004. "Reinventing Media Activism: Public Interest Advocacy in the Making of U.S. Communication-Information Policy, 1960–2002." *Information Society* 20(3): 169–187.

Mueller, Milton, Brenden Kuerbis, and Chrstiane Pagé. 2007. "Democratizing Global Communication? Global Civil Society and the Campaign for Communication Rights in the Information Society." *International Journal of Communication* 1: 267–296.

Norris, Philippa. 2001. *Digital Divide: Civic Engagement, Information Poverty and the Internet Worldwide*. Cambridge: Cambridge University Press.

———. 2004. "The Bridging and Bonding Role of Online Communities." In *Society Online: The Internet in Context*, ed. Philip N. Howards and Steve Jones, 31–42. Thousand Oaks, London, New Delhi: Sage Publications.

Onuf, Nicholas J. 1989. *World of Our Making: Rules and Rule in Social Theory and International Relations*. Columbia: University of South Carolina Press.

Ó Siochrú, Seán. 2004a. "Will the Real WSIS Please Stand up?" *Gazette: The International Journal of Communication* 66(3–4): 203–224.

———. 2004b. "Civil Society Participation in the WSIS Process: Promises and Reality." *Continuum: Journal of Media and Cultural Studies* 18(3): 330–344.

Ó Siochrú, Seán, and Bruce Girard. 2003. "Introduction." In *Communicating in the Information Society*, ed. Seán Ó Siochrú and Bruce Girard, 95–223. Gereva: UNRISD Publications.

Padovani, Claudia. 1993. "Il nuovo ordine mondiale dell'informazione e della comunicazione." *Pace, diritti dell'uomo, diritti dei popoli* 7(3): 109–124.

———. 2001. "Processi di globalizzazione: Democrazia, sovranità, comunicazioni e culture." In *Comunicazione globale: Democrazia, sovranità, culture*, ed. Claudia Padovani, 5–64. Torino: UTET.

———. 2004. "The World Summit on the Information Society. Setting the Communication Agenda for the 21st Century? An Ongoing Exercise." *Gazette, The International Journal of Communication* 66 (3–4): 187–191.

———. 2005a. "Debating Communication Imbalances from the MacBride Report to the World Summit on the Information Society: An Analysis of a Changing Discourse." *Global Media and Communication* 1(3): 316–338.

———. 2005b. "Civil Society Organizations beyond WSIS: Roles and Potential of a 'Young' Stakeholder." In *Visions in Process II—The World Summit on the Information Society*, ed. Olga Drossou and Heike Jensen, 37–45. Berlin: Heinrich Böll Foundation.

Padovani, Claudia, and Kaarle Nordenstreng. 2005. "From NWICO to WSIS: Another World Information and Communication Order?" *Global Media and Communication* 1(3): 264–272.

Padovani, Claudia, and Elena Pavan. 2007. "Diversity Reconsidered in a Global Multi-Stakeholder Environment: Insights from the Online World." In *The Power of Ideas: Internet Governance in a Global Multi-Stakeholder Environment*, ed. Wolfgang Kleinwächter, 99–109. Berlin: Germany Land of Ideas.

———. 2008. "Information Networks, Internet Governance and Innovation in World Politics." In *Electronic Constitution: Social, Cultural, and Political Implications*, ed. Francesco Amoretti, 154–173. Hershey, PA: IGI Global.

———. 2009a. "The Emerging Global Movement on Communication Rights: A New Stakeholder in Global Communication Governance? Converging at WSIS but Looking Beyond." In *Making Our Media: Global Initiatives toward a Democratic Public Sphere* (vol. 1), ed. Laura Stein, Dorothy Kidd, and Clemencia Rodriguez, 223–242. Cresskill, NJ: Hampton Press.

———. 2009b. "Fra reti tematiche e reti sociali: Un ritratto delle mobilitazioni sui diritti di comunicazione in Italia" *Quaderni di Sociologia* 53(49): 11–42.

———. 2011. "Actors and Interactions in Global Communication Governance: The Heuristic Potential of a Network Approach." In *The Handbook of Global Media and Communication Policy*, ed. Robin Mansell and Mark Raboy, 543–563. Oxford: Blackwell Publishing.

Padovani, Claudia, and Arjuna Tuzzi. 2004. "WSIS as a World of Words: Building a Common Vision of the Information Society?" *Continuum: Journal of Media and Society* 18(3): 360–379.

Papacharissi, Zizi. 2002. "The Virtual Sphere: The Internet as a Public Sphere." *New Media Society* 4(1): 9–27.

———. 2004. "Democracy Online: Civility, Politeness, and the Democratic Potential of Online Political Discussion Groups." *New Media Society* 6(2): 259–283.

Pappi, Franz U., and David Knoke. 1991. "Political Exchange in the German and the American Labor Policy Domain." In *Policy Networks: Empirical Evidence and Theoretical Considerations*, ed. Bernd Marin and Renate Mayntz, 179–208. Boulder, CO: Westview Press.

Paré, Daniel J. 2003. *Internet Governance in Transition: Who Is the Master of This Domain?* Lanham, MD: Rowman and Littlefield Publishers.

Pasquali, Antonio. 2003. "A Brief Descriptive Glossary of Communication and Information Aimed at Providing Clarification and Improving Mutual Understanding." In *Communicating in the Information Society*, ed. Seán Ó Siochrú and Bruce Girard, 95–223. Geneva: UNRISD Publications.

Pattberg, Phillip. 2006. *Global Governance: Reconstructing a Contested Social Science Concept. GARNET* working paper no.04/06, 2006. www.garnet-eu.org/fileadmin/documents/working_papers/0406.pdf.

Peak, Adam. 2004. "Internet Governance and the World Summit on the Information Society." Accessed August 2010. www.apc.org/en/pubs/issue/internet-governance-and-world-summit-information-s.

Peters, Guy B. 1995. "The Public Service, the Changing State, and Governance." In *Governance in a Changing Environment*, ed. Guy. B. Peters and Donald J. Savoie, 288–322. Montreal: McGill-Queen's University Press.

Peters, Guy B., and John Pierre. 1998. "Governance without Government? Rethinking Public Administration." *Journal of Public Administration Research and Theory* 8(2): 223–243.

Pickard, Victor. 2007. "Neoliberal Visions and Revisions in Global Communication Policy from NWICO to WSIS." *Journal of Communication Inquiry* 31(2): 118–139.

Polat, Raba K. 2005. "The Internet and Political Participation: Exploring the Explanatory Links." *European Journal of Communication* 20(4): 435–459.

Poster, Mark. 1995. "Cyberdemocracy: Internet and the Public Sphere." http://mydj.ru/docs/CyberDemocracy.doc.

Postmes, Tom, and Suzanne Brunsting. 2002. "Collective Action in the Age of Internet: Mass Communication and Online Mobilization." *Social Science Computer Review* 20(3): 290–301.

Powell, Walter W. 1990. "Neither Market nor Hierarchy: Network Forms of Organization." *Research in Organizational Behaviors* 12: 295–336.

Putnam, Robert. 2000. *Bowling Alone: The Collapse and Revival of American Community*. New York: Simon and Schuster.

Raboy, Mark. 2002, *Global Media Policy in the New Millenium*. London, UK: University of London Press.

Raboy, Mark. 2004. "The World Summit on the Information Society and Its Legacy for Global Governance." *Gazette: The International Journal for Communication Studies* 66(3–4): 225–232.

Raboy, Mark, and Claudia Padovani. 2010. "Mapping Global Media Policy: Concepts, Frameworks, Methods." *Communication, Culture and Critique* 3(2): 150–169.

Rhodes, R. A. W. 1996. "The New Governance: Governing without Government." *Political Studies* 44: 652–667.

——. 1997. *Understanding Governance: Policy Networks, Governance, Reflexivity, and Accountability*. Buckingham and Philadelphia: Open University Press.

Robertson, Roland. 1992. "Mapping the Global Condition." In *Globalization: Social Theory and Global Culture*, ed. Roland Robertson, 49–60. London, Thousand Oaks, New Delhi: Sage Publications.

Rogers, Richard. 2004. *Information Politics on the Web*. Cambridge, London: MIT Press.

——. 2006. "Mapping Web Space with the Issue Crawler." http://www.govcom.org/full_list.html.

——. 2008. "The Politics of the Web Space." www.govcom.org/full_list.html.

—— 2010. "Internet Research: The Question of Method." *Journal of Information Technology & Politics* 7: 241–46.

Rosenau, James N. 1995. "Governance and Democracy in a Globalizing World." In *Re-imagining Political Community: Studies in Cosmopolitan Democracy,* ed. Daniele Archibugi, David Held, and Martin Kohler, 28–57. Cambridge: Polity Press.

——. 1999. "Towards an Ontology for Global Governance." In *Approaches to Global Governance Theory*, ed. Martin Hewson and Timothy J. Sinclair, 287–302. Albany: State University of New York Press.

——. 2002. "Information Technologies and the Skills, Networks, and Structures That Sustain World Affairs." In *Information Technologies and Global Politics*, ed. James N. Rosenau and J. P. Singh, 275–288. Albany: State University of New York Press.

Sassen, Saskia. 2004. "Local Actors in Global Politics." *Current Sociology* 52(4): 649–670.

Schou, Arild. 1997. "Elite Identification in Collective Protest Movements: A Reconsideration of the Reputational Method with Application to the Palestinian Intifada." *Mobilization: An International Journal* 2(1): 71–86.

Schwartz, Joseph E. 1977. "An Examination of Concor and Related Methods for Blocking Sociometric Data." *Social Methodology* 8: 255–282.

Selian, Audrey N. 2004. "The World Summit on the Information Society and Civil Society Participation." *The Information Society* 20: 201–215.

Sikkink, Kathryn. 2002. "Restructuring World Politics: The Limits and Asymmetries of Soft Power." In *Restructuring World Politics: Transnational Social Movements,*

Networks and Norms, ed. Sanjeev Khagram, James V. Riker, and Kathryn Sikkink, 301–318. Minneapolis: University of Minnesota Press.

Simmel, Georg. 1955 [1908]. "The Web of Group Affiliations." In *Conflict and the Web of Group Affiliations,* trans. K. H. Wolff and R. Bendix, 125–195. New York: Free Press.

———. 1998 [1908]. *Sociologia.* Torino: Edizioni Comunità.

Simon, Leslie D., Javier Corrales, and Donald R. Wolfensberger. 2002. *Democracy and the Internet: Allies or Adversaries?* Washington, D.C.: Woodrow Wilson Center Press.

Singh, J. P. 2002. "Introduction: Information Technologies and the Changing Scope of Global Power and Governance." In *Information Technologies and Global Politics,* ed. James N. Rosenau and J. P. Singh, 1–38. Albany: State University of New York Press.

Slaughter, Anne-Marie. 2004. *A New World Order.* Princeton: Princeton University Press.

Snow, David, and Robert D. Benford. 1988. "Ideology, Frame Resonance, and Participant Mobilization." In *International Social Movement Research, Vol. I: From Structure to Action: Comparing Social Movement Research across Cultures,* ed. Bert Klandermans, Hanspeter Kriesi, and Sidney Tarrow, 197–217. Greenwich, London: Jai Press, 1988.

———. 1992. "Master Frames and Cycles of Protests." In *Frontiers in Social Movement Theory,* ed. A. D. Morris and C. M. Mueller, 135–155. New Haven: Yale University Press.

Snow, David, E. Burke Rochford, Steven Warden, and Robert D. Benford. 1986. "Frame Alignment Processes, Micromobilization and Movement Participation." *American Sociological Review* 51(4): 464–481.

Sørensen, Eva, and Jacob Torfing. 2007. "Introduction: Governance Network Research: Towards a Second Generation." In *Democratic Network Theories,* ed. Eva Sørensen and Jacob Torfing, 1–24. London: Palgrave Macmillan.

Steinberg, Marc W. 1998. "Tilting the Frame: Considerations on Collective Action Framing from a Discursive Turn." *Theory and Society* 27(6): 845–872.

Susskind, Lawrence E., Boyd W. Fuller, Michèle Ferenz, and David Fairman. 2003. "Multistakeholder Dialogue at the Global Scale." *International Negotiation* 8: 235–266.

Tannen, Deborah.1993. *Framing in Discourses.* New York: Oxford University Press.

Tannen, Deborah, and Cinthia Wallat. 1999. "Interactive Frames and Knowledge Schemas in Interaction: Examples from a Medical Examination/Interview." In *The Discourse Reader,* ed. Adam Jaworski and Nikolas Coupland, 332–348. London, New York: Routledge.

Tarrow, Sidney. 1994. *Power in Movement.* Cambridge: Cambridge University Press.

Ungar, Stu. 1992. "The Rise and (Relative) Decline of Global Warming as a Social Problem." *Sociological Quarterly* 33(4): 483–501.

United Nations High-Level Panel on Civil Society. 2004. "The Diversity of Actors within the UN System." www.un.org/reform/civilsociety/categories.shtml.

Uranga, Washington. 1986. "NWICO: New World Information and Communication Order." In *Communication for All*, ed. Philip Lee, 70–108. New York: Orbis Book.

Van Dijk, Jan. 1999. *The Network Society: Social Aspects of the New Media*. London: Sage.

Van Waarden, Frans. 1992. "Dimensions and Types of Policy Networks." *European Journal of Political Research* 21(1–2): 29–52.

Wasserman, Stanley, and Katherine Faust. 1994. *Social Network Analysis: Methods and Applications*. Cambridge: Cambridge University Press.

Wellman, Berry. 2002 [1988]. "Structural Analysis: From Method and Metaphor to Theory and Substance." In *Social Networks: Critical Concepts in Sociology* (vol. I), ed. John Scott, 70–108. London and New York: Routledge.

Wellman, Berry, Janet Salaff, Dimitrina Dimitrova, Laura Garton, Milena Gulia, and Caroline Haythornthwaite. 1996. "Computer Networks as Social Networks: Collaborative Work, Telework, and Virtual Community." *Annual Review of Sociology* 22: 213–238.

Wendt, Alexander. 1992. "Anarchy Is What States Make of It: The Social Construction of Power Politics." *International Organization* 46(2): 391–426.

———. 1999. *Social Theory of International Politics*. Cambridge: Cambridge University Press.

White, Robert A. 1986. "Christians Building a New Order of Communication." In *Communication for All*, ed. Philip Lee, 105–117. New York: Orbis Book.

DOCUMENTS AND RESOLUTIONS

International Telecommunication Union. "Resolution 1179/2001." www.itu.int/wsis/docs/background/resolutions/1179.html.

United Nations General Assembly. "Resolution 56/183." www.wsis-pct.org/resol-56–183.html.

Working Group on the Internet Governance. "Report on the work of Working Group on the Internet Governance." www.wgig.org/docs/WGIGREPORT.pdf.

WSIS. 2003. "WSIS Plan of Action." www.itu.int/wsis/docs/geneva/official/poa.html.

———. 2005. "Tunis Agenda for the Information Society." *WSIS-05/TUNIS/DOC/6(Rev. 1)-E*, 2005. www.itu.int/wsis/docs2/tunis/off/6rev1.html.

WSIS Civil Society. 2005. "Much More Could Have Been Achieved." www.csbureau.info/csdeclaration.htm.

Index

www.ingramcontent.com/pod-product-compliance
Lightning Source LLC
Chambersburg PA
CBHW071421050326
40689CB00010B/1931